AF863916

Urban and Landscape Perspectives

Volume 22

Series Editor

Giovanni Maciocco, Palazzo del Pou Salit, Università di Sassari, Alghero, Sassari, Italy

Urban and Landscape Perspectives is a series which aims at nurturing theoretic reflection on the city and the territory and working out and applying methods and techniques for improving our physical and social landscapes.

The main issue in the series is developed around the projectual dimension, with the objective of visualising both the city and the territory from a particular viewpoint, which singles out the territorial dimension as the city's space of communication and negotiation. The series will face emerging problems that characterise the dynamics of city development, like the new, fresh relations between urban societies and physical space, the right to the city, urban equity, the project for the physical city as a means to reveal civitas, signs of new social cohesiveness, the sense of contemporary public space and the sustainability of urban development. Concerned with advancing theories on the city, the series resolves to welcome articles that feature a pluralism of disciplinary contributions studying formal and informal practices on the project for the city and seeking conceptual and operative categories capable of understanding and facing the problems inherent in the profound transformations of contemporary urban landscapes.

More information about this series at http://www.springer.com/series/7906

Mauro Tebaldi • Marco Calaresu

The Quality of Territorial Policies in Europe's Periphery

Urban Regeneration and Environmental Protection

Springer

Mauro Tebaldi
Department of Law
University of Sassari
Sassari, Italy

Marco Calaresu
Department of Law
University of Sassari
Sassari, Italy

Urban and Landscape Perspectives
ISBN 978-3-030-62646-4 ISBN 978-3-030-62647-1 (eBook)
https://doi.org/10.1007/978-3-030-62647-1

© Springer Nature Switzerland AG 2021
This work is subject to copyright. All rights are reserved by the Publisher, whether the whole or part of the material is concerned, specifically the rights of translation, reprinting, reuse of illustrations, recitation, broadcasting, reproduction on microfilms or in any other physical way, and transmission or information storage and retrieval, electronic adaptation, computer software, or by similar or dissimilar methodology now known or hereafter developed.
The use of general descriptive names, registered names, trademarks, service marks, etc. in this publication does not imply, even in the absence of a specific statement, that such names are exempt from the relevant protective laws and regulations and therefore free for general use.
The publisher, the authors, and the editors are safe to assume that the advice and information in this book are believed to be true and accurate at the date of publication. Neither the publisher nor the authors or the editors give a warranty, expressed or implied, with respect to the material contained herein or for any errors or omissions that may have been made. The publisher remains neutral with regard to jurisdictional claims in published maps and institutional affiliations.

This Springer imprint is published by the registered company Springer Nature Switzerland AG
The registered company address is: Gewerbestrasse 11, 6330 Cham, Switzerland

Abstract This book focuses on territorial policies as instruments for local development in Europe's periphery. Using a multiple-case research design in three typical case studies in the context of the Mediterranean island of Sardinia (Italy), we empirically test the hypothesis that the institutionalisation of the governance system is an independent variable that is capable of influencing the quality of public policies, intended as a dependent variable. According to this hypothesis, the two variables mentioned above tend to change according to a linear and direct correlation: upward variation of the degree of institutionalisation of the governance system tends to correspond to upward variation in the quality of the policies, and vice versa. In our conclusions, we discuss the descriptive and prescriptive implications of the empirical findings of the research for the local development of peripheral/insular European contexts. Regarding the descriptive implications, we explain how territorial policy-making can be articulated and which schemes govern it, based on the degree of institutionalisation of the governance system and the quality of the territorial policies. Regarding the prescriptive implications, we identify the best practices for territorial governance to improve the chances of local development in Europe's periphery.

Keywords Quality of territorial policy · Institutionalisation · Urban regeneration · Environmental protection · Europe's periphery

Acknowledgements

This research—funded by the University of Sassari Research Fund 2019 (Fondo di Ateneo per la ricerca 2019)—is the result of a joint effort undertaken by the two authors who assume full responsibility for the contents of the volume as well as for any errors or omissions contained herein. All quotations originally published in languages other than English are translated by the authors. If some results have been achieved, this is thanks to our colleagues (in alphabetical order) Chiara De Micheli, Efisio Espa, Laura Lutzoni, Giovanni Maciocco, Andrea Migone, and Silvia Serreli, for discussing and debating ideas, concepts, and method related to this work. A special thanks go to Sara Aime, Stephen Bennetts, and Tonito Solinas for their tireless support in editing, translating, and proofreading the pages of this book since its very first draft. The authors would also like to thank the two anonymous referees for their constructive remarks and wish to pay a sincere tribute to the graduating students Stefano Erriu and Fabio Sanna at the University of Sassari for their hard work in conducting the interviews in the three typical case studies. A heartfelt thanks go to Gianni Calaresu, for using his photographic talent to bring the book cover to life. Finally, the authors wish to thank all the interviewees who have participated in this research: *grazie*!

Contents

List of Figures

List of Tables

Chapter 1
Conceptualising the Nexus Between Territorial Policies and Local Development in Europe's Periphery

Abstract This chapter aims to generalise the nexus between territorial policies and local development in Europe's periphery. First, we address the concept of periphery and how it can be analysed. Second, we explain why the Mediterranean islands can be considered Europe's periphery. Third, we present territorial policies and how they can be classified. Based on this conceptualisation, this chapter provides a brief description of the Italian political and administrative structure, followed by a review of the studies that have analysed the correlation between territorial policies and local development of peripheral areas in the Italian case. Therefore, we propose a series of preliminary questions, to steer territorial policies towards the local development of peripheral areas and reduce their risk of failure.

Keywords Territorial policies · Local development · Europe's periphery

1.1 The Concept of Periphery and How It Can Be Analysed

In the social sciences, the concept of periphery and the closely related concept of peripherality do not have a univocal meaning, since they refer to a complex, multifaceted phenomenon.[1] Therefore, to understand this concept from an analytical perspective and grasp all its nuances, a multidisciplinary approach is required. In addition, the composition of the facets of a periphery into a single multidimensional concept requires synthesis, which is the core objective of this section.

Since the early 1900s—but especially starting in the 1960s—the idea of dividing the world according to the concepts of 'centre' and 'periphery' has spread across disciplines, such as geography, political science, economics and sociology. Through

[1]In this regard, while exploring the concept of peripherality, Crone (2012, 50) describes it as 'a "fuzzy concept" since it arguably lacks clarity, is difficult to operationalise and possesses multiple meanings'.

© Springer Nature Switzerland AG 2021

M. Tebaldi, M. Calaresu, *The Quality of Territorial Policies in Europe's Periphery*, Urban and Landscape Perspectives 22,
https://doi.org/10.1007/978-3-030-62647-1_1

this process, these terms have acquired different meanings, partly depending on the subject-specific perspectives that are adopted in different scientific circles.

Human and economic geographic studies define periphery based on its spatial features. According to some authors, geographical features differentiate peripheral areas from central regions and regulate the conditions for the economic development—as well as the cultural and political development—enjoyed by the communities and businesses that populate the different regions. Knox and Marston (2001) interpret regional development from a spatial perspective, classifying regions as core, semi-peripheral or peripheral. Following their logic, the conditions that exert the greatest degree of influence on local economic development may be inferred based on the geographical location of each area (i.e. such development is higher in central areas and lower in peripheral areas). As a result:

> The core regions are dominated by trade, control of most advanced technologies and high labour productivity. The peripheral regions are characterised by dependent and disadvantageous business relations, more primitive technologies and closely specialised economy of lower productivity level. (Vaishar 2006, 7)

According to some scholars who studied issues connected to peripheral areas, the causes of peripherality and its characteristics cannot be explained using the spatial distance of a location from a hypothetical centre as the only parameter. For instance, in political science, the centre–periphery paradigm led to the development of a comprehensive 'territorial approach to politics'. This approach has been applied to many topics connected to politics (e.g. the creation of political parties and intergovernmental relations in democratic systems) based on a mutually recognised meaning of periphery. The periphery is conceptualised as the product of two interconnected dimensions that differ from an analytical point of view: the spatial and geographical horizontal dimension, which we referred to above, and the political, economic and cultural vertical dimension, which concerns the interactions between groups of stakeholders. From this perspective, the periphery is defined as a territorial area characterised by three key features: distance, difference and dependence. As Rokkan and Urwin (1983, 3) argue:

> A periphery is located at some distance from the dominant centre or centres, and its transactions with the latter are fraught with costs. A periphery is also different from the central areas on one or more scores: while the degree of distinctiveness will vary, being to some extent a function of distances and dependence, there will invariably be some minimum level and sense of separate identity. Finally, a periphery is dependent upon one or more centres in at least one of [the following] domains of behaviour [...]: in political decision-making, in cultural standardization, and in economic life.

A periphery may be seen then as a *structure of opportunities*, that is, a set of endogenous organisational, territorial, economic and institutional factors that offer several courses of action to inhabitants and impose quite a few constraints. However, the range and the attractiveness of these possibilities are affected by variables that are exogenous to the periphery. The most important of these exogenous variables are the resources and the political, economic and cultural strategies of the centre (Urwin 1991).

In developmental studies, dependency theory has conceptualised the centre–periphery dichotomy mostly from an economic point of view (Frank 1967; Wallerstein 1974; O'Brien 1982). This theory is based on analysis of unfair economic relations between strong centres, identified as the regions or countries with the highest industrial and infrastructural economic development, and the weak periphery, which relies on a mostly rural economy and is more disadvantaged due to the scarcity of infrastructure. It is an analytical perspective that tends to correlate the periphery with rural areas, although recently it has been applied to the study of specific economic activities, such as tourism, in insular areas that are also considered peripheral (Brown and Hall 2000). Regardless of the geographical scope, the theory attributes the origin of the unfair difference between centres and peripheries to the dependency of certain countries or regions on others. Such dependency is the unavoidable by-product of the dynamics and outcomes of capitalism, both nationally and internationally.

Within the economic development school of thought, the dependency theory has been challenged by other scientific approaches. In particular, two other theories with opposite explanations and solutions explore the nature and causes of the differences between centres and peripheries. The first is convergence theory, the prominent supporters of which were Friedmann (1966, 1973), Hirschman (1958, 1967) and Williamson (1965). Those authors argue that the economic–territorial difference between advantaged central areas and disadvantaged peripheral areas is a temporary, yet inevitable, phase of the economic development process. However, the effects of this phase are balanced by a series of factors that stimulate development in the peripheral areas and that, at a later stage of the process, tend to restore balance in the relationship between the centre and the periphery. The opposite view is supported by Myrdal (1957), one of the main representatives of the divergence school. This view:

> [. . .] holds that an equilibrium position between centre and periphery can never occur because of the over-riding mechanism of cumulative causation. Thus, the unavoidable persistence of certain immobilities (of power, population, natural resources etc.) continue to produce spread and backwash effects, making the rich, richer and the poor, poorer. (Oppong et al. 1988, 607)

As this brief overview of the literature shows, the attributes of the periphery and peripherality highlighted by the main theoretical approaches may be summarised as follows: (a) spatiality (the periphery is a geographically defined territory), (b) asymmetric relations (the definition of periphery implies a relationship with one or more centres that have a higher political, economic and social capacity), (c) identity (the periphery has specific social and cultural characteristics), (d) political nature (the periphery aims to influence decisions at various political and administrative levels), and (e) political, economic and social inequality and marginalisation (the asymmetrical relation with the centre demonstrates the uneven and marginal nature of peripheral areas compared to the central area, which often—but not necessarily—results in dependency).

Table 1.1 Classification of islands according to population

Category	Population size (permanent inhabitants)	Number of islands
Large islands	More than 50,000	15[a]
Medium-sized islands	5000–50,000	44
Small islands	50–4999	303
Very small islands	Less than 50	228

Source: Adapted from ESPON (2013)
[a]Of these, 5 have more than 500,000 inhabitants (Sicily, Sardinia, Mallorca, Cyprus and Crete)

In an effort to create a single multidimensional concept that includes all these attributes, we define a *periphery* as a *physical space* that is *territorially* and *culturally* different from one or more centres, with which it has an asymmetrical relation that affects the *political* choices it makes to balance its own political, economic and social *marginalisation conditions* (if not its dependency).

1.2 Why Mediterranean Islands Can Be Considered Europe's Periphery

The European Islands System of Links and Exchanges (EURISLES hereafter) network points out that all European Union (EU hereafter) islands are dealing with the same set of problems that Sardinia faced at the beginning of the last century: isolation and remoteness, limited natural and human resources, difficulties in terms of competitiveness and economic development and weak environments (EURISLES 1999, 2003). According to the most recent Programme for the Development of the Islands (ESPON hereafter), these islands represent territories that face the problems of permanently attracting active populations, being unable to create or attract competitive economic activities and having a limited socio-economic capability and divergence from the EU's objectives of sustainable development and economic, social and territorial cohesion (ESPON 2013).

Of the 362 European islands with a population of over 50 inhabitants (in addition to the 228 islands with a population of less than 50), 31 island regions (Table 1.1) are part of the European statistical units (Member States, Nomenclature of Territorial Units for Statistics [NUTS] II or NUTS III). Several types of common, fundamental data to empirically sustain the above-mentioned assumptions are available for these regions. These data can be divided into six categories: (a) population evolution, (b) economic efficiency, (c) economic activity of the population and unemployment rates, (d) functional urban areas (FUAs hereafter, see OECD 2013 for further definition), (e) educational level and (f) direct and indirect attractiveness indices.

(a) Regarding population evolution, although there are differences between islands in Europe, the demographic balance between deaths and births is generally negative. The decline of the young and working-age population is particularly significant. (b) Regarding economic efficiency, most of the islands show lower performance than their mainland counterparts, with an average per capita GDP of just 79.2% of that of the European market. (c) The data concerning the economic activity of the population show that unemployment rates are extremely high, particularly among youths and women, compared to the continental European average and national averages. (d) In addition, islands usually lack FUAs of local importance. Generally, insularity leads to small and isolated markets in which monopolies and oligopolies frequently replace the free market. In these contexts, for example, the prices of transportation and goods are higher compared to those in the rest of the European continent and the national reference economies. (e) Also, the educational level of islands is generally very low, even when at least public and/or private higher education institution (HEI) is present. (f) Finally, the index of direct and indirect attractiveness encompasses parameters conditioned by insularity, such as accessibility (from either a resident or non-resident viewpoint) and the agglomeration of economies. All the islands have accessibility indices lower than the European average, even without considering maritime transport. In addition, the costs of transport (in terms of time and expenditure as well as choice limitations) are higher on the islands than on the mainland.

1.3 Territorial Policies and How They Can Be Classified

Discussing territory or territoriality when considering public policies may seem obvious or even redundant since all public policies have their own (often implicit) territorial dimension. To confirm this, the definition of public policy developed by Dye (1992, 2)—'whatever governments choose to do or not to do'—can be useful. Since each government exercises sovereignty over a specific territory, public policies are inherently territorial, meaning that any authoritative decision, be it a 'do' or 'do not', generates a direct or indirect impact in the area where power is exercised. When the decision focuses on citizens or their movable assets, it indirectly affects the territory to varying degrees, depending on how citizens or their assets are present in it, cross it or generally benefit from it as an essential condition for their existence. In contrast, decisions directly affect when they specifically govern the transformation of the physical space in which citizens live or use to support their existence. This also applies to the immovable assets that the space hosts or provides to citizens. We can debate more specifically about territorial policies in relation to the second kind of decision since the scope of these decisions is the territory itself and citizens are involved based on the extent and scope of their engagement with the territory (Murphy 1989; Harris 2001; Governa and Salone 2004; Gualini 2004; Muro 2009).

To narrow the field of analysis, we follow Balme et al. (1994, 389), who define territorial policies as 'public programmes whose goal is the management of territories in their political and socio-economical dimensions'. Based on this definition, we highlight two specific features of territorial policies: their *location* (consequential from the management of territories) and *integration* (resulting from the need to harmonise the political and socio-economic dimensions).

Table 1.2 Types of territorial policies

		Localisation	
		Demarcation of the territorial decision-making sovereignty borders	*Definition of equality and freedom in the use of the territory*
Integration	*Horizontal*	Territorial governance (e.g. public–private partnership policies and network governance)	Territorial development (e.g. transport policy, infrastructure policy)
	Vertical	Governmental territorial policies (e.g. local government reform policy)	Territorial planning, use and transformation (e.g. urban policy, environmental policy)

Territorial policies are localised because they result from authoritative decisions that affect the transformation and conservation of the physical space of polities. Each polity, be it national, regional, local or supernational, is spatially limited. The physical space of polities defines the boundaries of decision-making sovereignty (i.e. the effectiveness of decisions and the political and administrative actors who have decision-making power) and restricts and localises the referent community, including its individual and collective issues, human resources and activities. In this process, it identifies the criteria for the actions aimed at regulating individual freedoms related to land use and lessening the differences arising from unequal use of the territory (De Decker et al. 2001; Camagni 2002; Rivolin 2005; Faludi 2007).

In this regard, the concept of localisation should not be confused with the concept of territorialisation. As emphasised by Storper (1997), the territorialisation of action simply means that the action occurs in a specific place, but when we argue about localisation, we refer to an action aimed at creating value in a specific place. In analytical terms, the requirement of localisation provides territorial policies with the following three characteristics: '[t]hey refer to the territory as a common good; they are concerned with the identification and valorisation of territorial capital; and the territory itself is defined during the action' (Davoudi et al. 2008, 44).

In addition to localisation, territorial policies involve integration. In the context of territorial policies, we can discuss integration in two ways: either horizontally, by considering the dimensions (political, social and economic) affected by the decisions governing the physical space of polities, or vertically, by considering the jurisdictional levels that necessarily interact with territorial decision-making processes and, for this reason, require integration and coordination.

Analytically speaking, the territorial policy is an extensive category that may include different public policies with the characteristics described above. A preliminary classification (Table 1.2) may be developed based on these characteristics.

By combining integration and localisation criteria, four kinds of territorial policies may be defined. When localisation is intended to define the boundaries within which decision-making sovereignty is exercised, and integration is horizontal, territorial governance policies are applied. This category includes policies that establish partnerships between public and private stakeholders to manage specific policy areas and policies that govern the application of governance network models. The same localisation principle combined with vertical integration results in governmental territorial policies, such as local and regional government reform policies.

Localisation may also be seen as a guiding principle to reduce social inequality in the use of public spaces. Combining with a horizontal integration criterion can result in territorial development policies, such as transport policies and infrastructure policies. If combined with vertical integration, the resulting policies concern the planning and regulation of territorial use and transformation, such as town planning and environmental policies.

For reasons that are explained in detail in the next chapter (Chap. 2), our empirical research examines a specific type of territorial policy—which includes planning, land-use regulation and territorial transformation—and analyses urban regeneration and environmental protection policies.

1.4 Territorial Policies and Local Development in Europe's Periphery: An Assessment of the Italian Case

The scientific literature exploring the links between regional policies and the development of the European peripheral areas is rich and varied. Despite their diverse purposes and guidelines, research efforts seem to have one trait in common: the hypothesis that territorial policies should foster local development in peripheral areas and all 'fragile areas' in general.[2] Without going into the controversial interpretation of the concept of 'local development' (Moulaert and Nussbaumer 2008; Moulaert and Mehmood 2010; Pike et al. 2010), we instead consider a broad definition that identifies local development as the result of both the economic and socio-political dimensions (Trigilia 2001, 2005). Implicitly and theoretically, we assume that local development is 'deeply historical, place-specific and embedded within specific and concrete institutional settings' (Moulaert et al. 2007, 196).

From this perspective, local development takes place when the following occurs in an area, generally a group of non-metropolitan municipalities, metropolitan areas or provincial areas (but not necessarily the Italian provincial administrative boundaries). First, from an economic point of view, local development occurs when there is a significant and sustained increase in investments, especially in private

[2]Following Zanon (2014, 343–344), a 'fragile area' can be defined 'on the basis of recent studies (Governa 2008; Corrado 2010) and further analysis in terms of a combination of critical conditions and unique values—often at risk: complex morphology, hydrogeological problems/natural values and scenic beauty; depopulation, weakness of familial and community networks/local cultures, and local knowledge; lack of urban facilities and services/quality of traditional settlements; economic change, weakness of local resource-based sectors, and small size of enterprises/niche products, traditional products; change of scale of spatial, economic and social relationships/permanence of proximity ties and relationships; area set apart, "dumping site" for unwanted urban activities and playground for urban users/preserved spaces'.

investments, production facilities, production (both physical and in terms of monetary value), exports, productivity and employment. Second, at the socio-political level, local development occurs when there is a similar increase in social capital[3] among community members and the 'institutional thickness'[4] supporting the actual implementation of the development process (La Spina 2008; Zanon 2014).

Based on this conceptualisation, this section provides a brief description of the Italian political and administrative structure, followed by a review of the studies that have analysed the correlation between territorial policies and local development of peripheral areas in the Italian case. We limit our analytical focus to the Italian case because the scope of the empirical research concerns the Italian insular territory of Sardinia, which we have already defined as a peripheral, underdeveloped territory, and the Italian case is typical within the European context. As Governa and Salone claim (2004, 800):

> In the Italian context, changes in territorial organization and their connection to urban and territorial policies present a number of particularly interesting elements. These may be identified by focusing attention on the changes resulting from the construction of territorial transformation processes and on the innovations introduced by 'new' instruments for the intervention in the city and in the territory.

1.4.1 *The Italian Political and Administrative Structure*

To describe the institutional context within which the territorial policies were adopted, we first provide some basic information about the Italian political and administrative structure. The Italian Republic can be defined as a 'multi-level polity' (Longo and Mobilio 2016, 2) which, according to Art. 114 of the Constitution, is currently made up of municipalities (*Comuni*), provinces (*Province*), metropolitan cities (*Città metropolitane*), regions (*Regioni*) and the state (Baccetti, 2008).[5]

---

[3]In the social sciences, the concept of social capital has different meanings. In this volume, we use the definition developed by Putnam (1993, 167), according to which social capital refers to 'features of social organization, such as trust, norms, and networks, that can improve the efficiency of society by facilitating coordinated actions'.

[4]The concept of 'institutional thickness' developed by Amin and Thrift (1995) concerns the development of specific institutional features in each area, which can be summarised as follows: (a) a strong institutional presence, (b) a high level of interaction among these institutions to facilitate reflexive networking, cooperation and information exchange; (c) well-defined structures of domination, coalition-building and collective representation in order to minimise sectionalism and rogue behaviour; and (d) the emergence of cognitive mapping of places that leads agents to perceive a common agenda upon which the collectivisation of institutions depends and develops (MacLeod 1997).

[5]As stated by Baccetti (2011, 164), it is worth noticing that 'the Italian words "Regione", "Provincia" and "Comune", unlike their English equivalent, refer to both the relevant geographical area and to the local government body associated with that geographical area'.

The regions (*Regioni*) are the first subnational unit of division of the country, constituting Italy's second NUTS[6] administrative level (Eurostat 2011). As defined by the Constitution (Art. 123), each region adopts its own statute through constitutional laws, which determines the form of government, the fundamental principles of its organisation and the functioning of the region. There are currently 20 regions in Italy. Of these, five (*Regioni a Statuto Speciale*) are constitutionally granted extended autonomy by special statutes, and the rest (*Regioni a Statuto Ordinario*) are governed by an ordinary statute (Bolgherini 2014). The legislative powers of both special and ordinary regions are subject to certain constitutional limitations, the most important of which is that regional acts may not conflict with national interests.

Apart from Valle d'Aosta,[7] each region in Italy is divided into provinces (*Province*), which form the intermediate- or second-tier territorial level and correspond to Italy's third NUTS administrative level (Eurostat 2011). Provinces in Italy, at least before the 2014 reform (Law 56/2014) and the introduction of metropolitan cities,[8] represented 'the middle layer in the government structure, both in terms of political devolution and administrative decentralisation' (Baccetti 2011, 165). Currently, there are 110 provinces in Italy (Bolgherini 2014). A province has statutory autonomy within the limits set by the state and regional laws, with major responsibilities related to the environment and town and rural planning (Baccetti 2011).[9] In each province, there is also a prefect (*Prefetto*) chairing a prefecture (*Prefettura*), who serves as the representative of the central state authority for public safety and law enforcement.

Municipalities (*Comuni*) are the basic subnational unit of the state, roughly equivalent to a township or a city, and they comprise Italy's second local administrative unit (LAU) level (Eurostat 2011). There are over 8000 municipalities in Italy (Bolgherini 2014). They have the power to levy and collect limited local taxes, and they have their own police forces to manage public safety and urban security (*Polizia Municipale e Locale*). However, the powers of these police forces are much more limited compared to the powers exercised by the national police forces (*Polizia di Stato*, *Arma dei Carabinieri* and *Guardia di Finanza*).[10]

---

[6]The NUTS provides a uniform, consistent categorisation of territorial units to produce regional statistics for the European Union (Eurostat 2011).

[7]The Valle d'Aosta region is the sole exception: it is not divided into provinces, and provincial functions are exercised by the region.

[8]Although this is beyond the scope of this book, we can say that the metropolitan cities in Italy are hybrid administrative entities within the Italian local governmental system that represent both a metropolitan community, which is similar to the old provincial community and the various municipalities located within the boundaries of the metropolitan area (Longo and Mobilio 2016).

[9]South Tyrol and Trentino-Alto Adige are autonomous provinces. Unlike all other provinces in Italy, they have the same legislative powers retained by the regions, and they are not subject to the power of the region in which they are located, namely, Trentino-Alto Adige/Südtirol.

[10]See Calaresu and Tebaldi (2020) for a description of the Italian national police system in comparative perspective.

1.4.2 Evaluation Studies of Territorial Policies and Local Development in Italy's Periphery

During the 1990s and the first decade of the twenty-first century, several innovative territorial programmes were launched in Italy in order to strengthen local development in the weaker areas of the country. Those innovative programmes, which take the form of contracts and pacts[11] can be divided into two groups (Governa and Salone 2002). The first group includes 'complex urban programmes promoted by the Ministry of the Public Works based on a number of contractual urban planning experiments conducted in several regional contexts' (Governa and Salone 2004, 804).[12] The second group consists of six instruments for negotiated programming promoted by the Ministry for the Treasury, Budget and Economic Planning during the 1990s[13] (Governa and Salone 2002; La Spina 2008). These instruments were originally conceived as instruments to promote development in southern Italy and:

> [. . .] then became the methodology applied in promoting development throughout the country, thus facilitating Italian instrumentalization of requests from EU institutions [. . .] Possibilities for intervention throughout Italy were thus extended, particularly in areas falling under Objectives 1 (underdeveloped areas), 2 (areas in industrial decline) and 5b (fragile rural areas) as defined by the EU. (Governa and Salone 2004, 805)

Evaluation studies that have analysed the correlation between territorial policies and the local development of Italian peripheral areas have examined some specific forms of territorial policies, especially the two most common forms, Territorial Pacts and Integrated Territorial Programmes.

Studies that limited their analyses to the output production phase only registered a quantitative increase in outputs. According to Governa and Salone (2004, 810), 'territorial pacts have been promoted in nearly all the Italian Regions, although there

---

[11]Contracts and pacts are intended here as a form of public agreement in which the actors involved declare to approve a project or a line of action, taking mutual commitments, making their resources (not necessarily financial resources) available for common action and agreeing how and when to act (Bobbio 2000). Using this kind of public agreements as a governance instrument, that is 'governing by contract' (Gaudin 1999; Cooper and Cooper 2002; Lascoumes and Le Galès 2004), would be in many occasions a response to the need of reconfiguring the relationships—and the responsibilities—between the national and the local level of government inside a state (see Calaresu 2017 and Calaresu and Triventi 2019) as an example of 'governing by contract' in the field of urban security). In this perspective, the 'spread of contractual practices' has been considered 'one of the most significant processes of political-administrative change' (Bobbio 2000, 112) in the last decades.

[12]These programmes are *Programmi di Recupero Urbano* (Urban Recovery Plans), *Programmi di Riqualificazione Urbana* (Urban Regeneration Programmes), *Contratti di Quartiere* (Neighbourhood Contracts) and *Programmi di Riqualificazione Urbana e di Sviluppo Sostenibile del Territorio* (Urban Regeneration and Sustainable Territorial Development Programmes).

[13]These negotiated programming tools are *Intese Istituzionali di Programma* (Institutional Programme Agreements), *Accordi di Programma Quadro* (Framework Programme Agreements), *Contratti di Programma* (Programme Contracts), *Contratti d'Area* (Area Contracts), *Patti Terrioriali* (Territorial Pacts) and *Programmi Integrati Territoriali* (Integrated Territorial Programmes).

is a clear concentration of initiatives in the traditionally weaker areas of the country—that is, the Regions of southern Italy'. However, this is a very one-dimensional analysis, since it is limited to quantitative evaluation of the output and can provide no information on the results or impact of implemented policies.

In terms of evaluation, the research conducted by Governa (2008) explores the implementation of integrated local development programmes in some peripheral and fragile areas in the mountain areas of Lombardy (Objective 2 areas). The territorial context in which such action programmes are expected to stimulate the development process is affected by social issues at both the political and administrative levels. As Governa (2008, 7) puts it:

> [the challenge is] that of promoting local development in the fragile areas of a strong Region, valorising the endogenous resources of the mountains, building partnership among tired and poorly motivated actors and activating the planning capability of a society that tends to be aphasic.

Most of the processes of implementing territorial policy highlight a number of critical issues, especially regarding the weak integration of stakeholders and the territories in which the specific programmes operate. Additionally, from a wider geographical perspective, the processes of implementing territorial policy highlight issues regarding the integration of different action programmes implemented in neighbouring territories. This weakness primarily concerns the architecture of the governance designed and implemented to execute territorial policies in situations when empirical analysis detected minor or no institutionalisation of vertical and horizontal integration mechanisms. The weak development of vertical integration mechanisms is especially important since:

> [...] these vertical governance mechanisms give access to those resources, first of all cognitive, through which to consolidate the capability of the administrations to reproduce the social capital and activate the planning capability of local society, which it is unable to express alone. Thus, forming social and territorial ties, but also and above all promoting co-operation among enterprises, social actors and public administrations. (Governa 2008, 11)

Although it highlights the strengths and weaknesses of the territorial policy governance model and any corrections that are necessary to obtain desirable effects in terms of local development, ongoing analysis of implementation is of little use for understanding whether the positive correlation between regional policies and local development exists in the real world and, most importantly, under which conditions. The most convincing step in this direction is taken by scholars (De Vivo 2004; Cerase 2005; La Spina 2008; Zanon 2014) who perform an ex post analysis of the results of territorial policies and assess the results based on the level of local development that they achieve in peripheral regions.

In general, evaluation studies of the achievements of Territorial Pacts and/or Territorial Integrated Programmes in terms of local development in peripheral and inherently fragile areas show concordant results. The results can be evaluated as satisfactory in territories where, despite the absence of historically and traditionally consistent social capital or the presence of a share of historical social capital that remains latent, the absent/latent social capital is created/activated and then implemented in communities via territorial policies. The case study analysed by Zanon (2014) focuses on some alpine areas of the Trentino region in northern Italy, which qualify as peripheral and fragile areas based on their social, economic and territorial profiles. In these areas were adopted territorial policies (specifically, Territorial Pacts) designed to stimulate local development. According to Zanon (2014), the adopted policies are successful because they primarily strengthened the social capital that was already present but latent in those areas. This stimulated re-territorialisation processes, which both had a multiplying effect on the reactivated social capital and fostered dynamism and economic development. At the end of his discussion, Zanon (2014, 353) confirms that:

> [...] it is possible to start successful initiatives in marginal places, where there is a dearth of economic capital, and physically embedded capital is weak. Social capital can be the key asset, but it cannot be simply understood as a heritage from the past or a generic disposition to cooperate with the local society; rather, it is a social construct which needs operational and feasible objectives able to involve various actors pursuing personal interests which are not in conflict with those of others.

La Spina (2008) provides empirical evidence of the correlation between territorial policies and local development in peripheral areas and, unlike the previous study, analyses failures. According to La Spina himself (2008), this correlation is confirmed by the case studies, which focus on the evaluation of Territorial Pacts and Territorial Integrated Programmes in the Campania and Sicily regions in Italy (De Vivo 2004; Cerase 2005). Both cases clearly show the ineffectiveness of the implementation of regional policies through these instruments when the basic social components of local development are not fostered. This analysis supports the hypothesis that when the social capital of a given community is not consistently nurtured, its ability to achieve a certain degree of economic development is considerably reduced. In the Campania and Sicily regions, where the limited pre-existing social capital (Putnam 1993) was not improved by territorial policies, Territorial Pacts were thought to be unsuccessful in terms of their ability to foster economic growth and employment and inefficient in terms of spending capacity (La Spina 2008).

1.5 Territorial Policies as Instruments for the Local Development of Peripheral Areas: Preliminary Research Questions

Based on our review of studies focusing on the connection between territorial policies and local development in Italian peripheral areas, we can identify five background assumptions:

1. Firstly, not all territorial policies bring about improvements in the local development of peripheral areas. Some policies have positive effects, while others have

no or adverse effects on the target social groups. Based on the above-mentioned evaluation of these policies, the positive correlation of the two variables is determined by social and historical factors that create the social capital of a given community. The relationship between these variables logically means that a remote explanatory factor (a social–historical event which took place at a certain point in the past) is able to determine an intermediate explanatory factor (the level of the social capital), which in turn determines a further intermediate political–institutional factor (the level of efficiency/effectiveness of territorial policies), which finally will determine the outcome of the phenomenon in question (the level of local development of peripheral areas).

2. Secondly, although the theory of social capital has been applied both in a comparative perspective and to various case studies (including the Italian case) and numerous empirical studies have confirmed its validity, the academic debate on the complex relationship between territorial policies and local development has highlighted some theoretical weaknesses. Numerous authors underline the existence of other remote explanatory factors not considered by social capital theorists. Proponents of *modernisation* (e.g. Rostow 1960, Hoogvelt 1976, Roxborough 1979, Bull and Bøås 2012) and *convergence* (e.g. Hirschman 1958, 1967, Williamson 1965, Friedmann 1966, 1973) theories maintain that it is economic–infrastructural factors which determine the long-term processes of development, both on a national and local level. In these theories, the state is considered a key factor in defining economic policies designed to support and promote processes of growth, quite independently of specific localities. Territorial policies are said only to be capable of creating economic development if an appropriate economic and infrastructural context also supports them. In contrast, adherents of the *territorial* approach (e.g. Palermo 2004; Magnaghi 2010, 2012; Governa 2014) concentrate their attention on the role of cultures, traditions and values, which are considered a product of relationships between individuals and the local territory in any given context. Geographical space is understood as a territorial area which, on the one hand, is determined by a historical relationship with its inhabitants who over time have modified and shaped it but which, on the other hand, also conditions the development of social groups and economic forces through the cultures, traditions and values maintained in the same territory. Those scholars who follow this approach, therefore, focus on the long-term influence which the geographical space tends to exercise over the values, cultures and traditions shared by human groups; the latter in turn influence the policy decisions which are made and, in particular, policy decisions directed towards local development. Other scholars of the sociological school underline the close link between social and economic structures, in a research stream that considers economic and political action as socially embedded (Granovetter 1985; De Janvry and Sadoulet 2007; Lazzeroni et al. 2013). According to promoters of this theory, it is long-term relational-type factors—or in other words, the structuration of relational networks between individuals—which determine the cultural characteristics and value systems of the community itself. These characteristics in turn, condition the process of political and social production,

including territorial policies, the effects of which in turn influence local development. Apart from those authors who challenge the social capital hypothesis on a theoretical level, other authors limit themselves, based on empirical studies, solely to highlighting the presence of significant exceptions. What seems particularly significant in this context are studies identifying the weak or non-existent correlation between the high level of social capital present in certain Italian provinces and the incidence of inefficient or ineffective territorial policies within these same areas. One example is the case of the Province of Rovigo in the Veneto Region. As Cabria states in a recent empirical study, 'the level of social capital in the Rovigno area [. . .] is particularly high, the institutional infrastructure of the area cannot be considered different from the surrounding regional context, nor is there a historically consistent organised crime in the area which might have conditioned the progress of economic development throughout the history of the province' (Cabria 2020, 38). According to an analysis by the same author, 'there is a weakness at the local level that we could define as historical and which has long impeded change. This factor has influenced economic development and the social fabric of the entire local system' (Cabria 2020, 32).

3. Thirdly, in contrast to the research studies mentioned above, we believe that the progress of research on the relationship between territorial policies and local development needs to concentrate, above all and specifically, on an intermediate theoretical level. This means setting aside remote-type explanatory factors—whether historical, economic, political and/or social—and focusing on intermediate-level empirical studies, based on the relationship between other explanatory elements than social capital, which seem able to clarify how and why territorial policies can produce local development outcomes.
4. Fourthly, in order to further clarify what we intend to analyse here and what will instead remain outside this book, we can state that the specific object of this research is the analysis of the political–institutional factors which we think have the greatest influence on the production of territorial policies and the quality of such policies. Therefore, this study will not consider (except incidentally) long-term historical, political, economic and social factors that may influence the politico–institutional independent variables that are the focus of our analysis. Nor will this study evaluate the concrete outcomes of the policies analysed in terms of local development.
5. Fifthly, in order to better specify our research approach, we can state that for real improvement to take place in the local development of peripheral areas, territorial policies must be of good quality. Therefore, the principal research points we aim to address in this book is understanding how good quality territorial policies can be developed, what factors foster such policies and how they can be introduced and replicated at the political and institutional levels.

There is general agreement in the social and political science literature that public policies tend to have consequences for communities of individuals. Once implemented, whether we consider their outcomes or impacts, policies generate more or less significant effects on societies (Brewer and deLeon 1983; deLeon 1988). However, such consequences do not always meet the expectations of policy-makers, as the results or impacts of public policies (a) may worsen, rather than improve, the social issues that require public intervention, (b) may affect a social group other than the targeted one or (c) may generate other issues in different sectors of society and/or contexts in which citizens operate (Ingram and Mann 1980). Our research is developed to precisely understand the weaknesses and strengths of public programmes regarding their expected results for specific groups. This can help minimise the margin of error or incidence of unexpected factors (Nachmias 1979).

One of the main questions that scholars must try to answer concerns the development of theories (Sabatier 1999), policy instruments (Linder and Peters 1989) and good quality action programmes or public policies that (a) can improve pre-existing social conditions, (b) mainly affect the expected social target groups and (c) do not cause additional problems for other social groups and/or their environment. These same issues apply when scholars and policy-makers consider territorial policies and seek solutions for problems connected to the weak local development of peripheral areas, such as island regions. We, therefore, propose a series of preliminary questions, in order to steer territorial policies towards the local development of peripheral areas and reduce their risk of failure. These questions are as follows:

(a) *What are good quality policies, and how can be they analysed?*
(b) *What political and institutional factors can be identified as specific variables which could explain the production of good quality territorial policies?*
(c) *Is it possible to develop a theoretical pathway and a subsequent empirical test that specifies which features should be included in territorial policies to ensure that they are considered good quality policies?*

1.6 Outline of the Book

This volume focuses on territorial policies as instruments for local development in Europe's periphery. In particular, it aims to empirically test the nexus between the institutionalisation of the governance system and the quality of urban regeneration and environmental protection policies in three typical case studies in the Mediterranean island of Sardinia, Italy.

The first part of the volume (Chap. 2) aims to develop the theoretical and methodological frameworks of the research. In the theoretical section, we clarify the explanatory objectives of the study. First, we explain the nature of public policies and how they can be analysed. Second, we introduce the institutionalisation of the governance system as an independent variable that can provide descriptive and prescriptive explanations of the quality of territorial policies from both politics- and policy-oriented perspectives. We then present a unified policy-oriented theoretical framework, explaining the analytical dimensions of both the above-mentioned variables and the relationship between them. The following theoretical hypothesis is proposed: there is a cause–effect relationship between the degree of institutionalisation and the outcomes of territorial policies regarding their capacity to respond to the issues affecting peripheral areas. In the methodological section of the chapter, we first address the selection of the island of Sardinia as the territorial context of the research. Second, we present the multiple-case research design and explain why urban regeneration and environmental protection were chosen to address the policy areas in the study. Finally, we discuss the choice of semi-structured interviews as the qualitative method of analysis to empirically test the research hypothesis.

The second part of the volume (Chaps. 3 and 4), which is based on the theoretical and methodological framework developed in the previous chapter, is devoted to the empirical analysis of the selected policy areas and typical cases. For each policy area (urban regeneration and environmental protection policy), we assess one or more case studies with different time spans (1995–2006 for the typical case study of the cities of Sassari and Alghero and 1993–2004 for the typical case study of the *Area Marina Protetta della Penisola del Sinis – Isola di Mal di Ventre* (Sinis peninsula – Mal di Ventre Island MPA hereafter). These two chapters are further divided into three sections each. In the first section of each chapter, we present the urban regeneration and environmental policies in Italy and Sardinia, addressing the relevant actors, objectives, resources, and instruments embedded in each policy. In the second section, we test the degree of institutionalisation and the quality of the policy in each policy area within the selected cases. For this purpose, we highlight several analytical dimensions of the institutionalisation of the governance system as well as the quality of the territorial policies. In the third section, we perform an overall assessment of the case studies in terms of each policy area, paying particular attention to evaluating the institutionalisation of the governance system and the quality of each observed territorial policy.

The third and final part of the volume (Chap. 5) is devoted to generalising the empirical evidence obtained in the previous chapters. Starting with a study of the policy areas of urban regeneration and environmental protection in the selected Sardinian case studies, we discuss the descriptive and prescriptive implications for the local development of other peripheral/insular European contexts. Regarding the descriptive implications, in the first part of the chapter, we explain how territorial policy-making can be articulated and which schemes govern it, based on the degree of institutionalisation of the governance system and the quality of the policy for each policy area. Regarding the prescriptive implications, in the second part of the chapter, we identify the best practices for territorial governance in order to improve the chances of local development in Europe's periphery.

References

Amin A, Thrift N (1995) Globalisation, institutional 'thickness' and the local economy. In: Healey P (ed) Managing cities. The new urban context. Wiley, Chichester, pp 91–108

Baccetti C (2008) La nuova politica locale. UTET, Torino

Baccetti C (2011) Italy. In: Heinelt H, Bertrana X (eds) The second tier of local government in Europe. Routledge, London/New York, pp 164–184

Balme R, Garraud P, Hoffmann-Martinot V, Le May S, Ritaine E (1994) Analysing territorial policies in Western Europe: the case of France, Germany, Italy, and Spain. Eur J Polit Res 25(4):389–411

Bobbio L (2000) Produzione di politiche a mezzo di contratti nella pubblica amministrazione italiana. Stato & Mercato 20(1):111–142

Bolgherini S (2014) Can austerity lead to recentralisation? Italian local government during the economic crisis. South Eur Soc Polit 19(2):193–214

Brewer G, deLeon P (1983) The foundations of policy analysis. The Dorsey Press, Homewood

Brown F, Hall D (eds) (2000) Tourism in peripheral areas. Case studies. Channel View Publications, Clevedon

Bull B, Bøås M (2012) Between ruptures and continuity: modernisation, dependency and the evolution of development theory. Forum Dev Stud 39(3):319–336

Cabria M (2020) La spirale della marginalità. Il caso del Polesine tra intervento pubblico e potere locale. Unpublished, Ph.D. thesis

Calaresu M (2017) Security pacts: the Italian experience. How contracts are used for governing crime and disorder. Eleven International Publishing, The Hague

Calaresu M, Triventi M (2019) Governing by contract as a way to reduce crime? An impact evaluation of the large-scale policy of security pacts. Policy Sci 52(2):255–279

Calaresu M, Tebaldi M (2020) The comparative analysis of European national police systems: a theoretical and methodological proposal for a new research agenda. Police Pract Res 21(6):640–654

Camagni R (2002) On the concept of territorial competitiveness: sound or misleading? Urban Stud 39(13):2395–2411

Cerase FP (ed) (2005) Azione pubblica e imprenditorialità. L'esperienza dei patti territoriali in Campania. Angeli, Milano

Cooper PJ, Cooper P (2002) Governing by contract: challenges and opportunities for public managers. CQ Press, Washington, DC

Corrado F (2010) Fragile areas in the Alpine region: a reading between innovation and marginality. Revue de Géographie Alpine 98(3)

Crone M (2012) Re-thinking 'peripherality' in a knowledge-intensive service-dominated economy. In: Danson M, De Souza P (eds) Regional development in Northern Europe: peripherality, marginality and border issues. Routledge, London, pp 49–64

Davoudi S, Evans N, Governa F, Santangelo M (2008) Territorial governance in the making. Approaches, methodologies, practices. Boletín de la Asociación de Geógrafos Españoles, 46:33–52

De Decker P, Meert H, Peleman K (2001) Combatting urban poverty. The power and specificity of territorial policies. In: De Decker P, Meert H, Peleman K (eds) Conference on "social inequality, redistributive justice and the City". International Sociological Association, Research Committee on Urban and Regional Development, Amsterdam, pp 15–17

De Janvry A, Sadoulet E (2007) Toward a territorial approach to rural development. *eJADE* 4 (853-2016-56115):66–98

De Vivo P (2004) Pratiche di concertazione e sviluppo locale. L'esperienza dei patti territoriali e dei Pit nella regione Campania. Angeli, Milano

deLeon P (1988) Advice and consent: the development of the policy sciences. Russell Sage Foundation, New York

Dye TR (1992) Understanding public policy. Englewood Cliffs, Prentice-Hall

ESPON (2013) Programme for the development of the islands. European Islands and Cohesion Policy (EUROISLANDS), ESPON & University of the Aegean, https://www.espon.eu/export/sites/default/Documents/Projects/TargetedAnalyses/EUROISLANDS/FinalReport_foreword_CU-16-11-2011.pdf

EURISLES (1999) European Islands system of links and exchanges. The island regions and the price of intra-EU transport of goods, https://europeansmallislands.com/tag/eurisles/

EURISLES (2003) European Islands System of Links and Exchanges. Islands transport and the EU, the situation in 2003. Conference of Peripheral Maritime Regions (CPMR) Islands Commission, https://europeansmallislands.com/tag/eurisles/

Eurostat (2011) Nomenclature of territorial units for statistics (NUTS), https://ec.europa.eu/eurostat/web/products-manuals-and-guidelines/-/KS-RA-11-011

Faludi A (2007) Territorial cohesion policy and the European model of society. Eur Plan Stud 15 (4):567–583

Frank AG (1967) Capitalism and underdevelopment in Latin America. Monthly Review Press, New York

Friedmann J (1966) Regional development policy. The MIT Press, Cambridge, MA

Friedmann J (1973) Urbanization, planning and National Development. Sage, Beverly Hills

Gaudin JP (1999) Gouverner par contrat. L'action publique en question. Presses de SciencesPo, Paris

Governa F (2008) Local development in fragile areas. Critical remarks on the initiatives underway in the mountains of Lombardy. J Alp Res 96(3):1–15

Governa F (2014) Tra geografia e politiche. Ripensare lo sviluppo locale. Donzelli Editore, Roma

Governa F, Salone C (2002) Descrivere la governance. Conoscenza geografica e modelli di azione collettiva nelle politiche urbane e territoriali. Bollettino della Società Geografica Italiana XII 7 (1):29–50

Governa F, Salone C (2004) Territories in action, territories for action: the territorial dimension of Italian local development policies. Int J Urban Reg Res 28(4):796–818

Granovetter M (1985) Economic action and social structure: the problem of embeddedness. Am J Sociol 91(3):481–510

Gualini E (2004) Regionalization as 'experimental regionalism': the rescaling of territorial policy-making in Germany. Int J Urban Reg Res 28(2):329–353

Harris N (2001) Spatial development policies and territorial governance in an era of globalization and localization. In: Organisation for Economic Co-operation and Development. (OECD) (ed) Towards a new role for spatial planning. OECD, Paris, pp 33–58

Hirschman AO (1958) The strategy of economic development. Yale University Press, New Haven

Hirschman AO (1967) Development projects observed. The Brookings Institution, Washington, DC

Hoogvelt AM (1976) The sociology of developing societies. Macmillan, London

Ingram HM, Mann DE (eds) (1980) Why policies succeed or fail. Sage, Beverly Hills

Knox PL, Marston SA (2001) Human geography: places and regions in a global context. Pearson, London

La Spina A (2008) Ambitions and failures of some local development policies for the Mezzogiorno. Territorial agreements and integrated territorial programs. Quaderni di Sociologia 48:21–38

Lascoumes P, Le Galès P (eds) (2004) Gouverner par les instruments. Presses de Sciences Po, Paris

Lazzeroni M, Bellini N, Cortesi G, Loffredo A (2013) The territorial approach to cultural economy: new opportunities for the development of small towns. Eur Plan Stud 21(4):452–472

Linder SH, Peters BG (1989) Instruments of government: perceptions and contexts. J Publ Policy 9 (1):35–58

Longo E, Mobilio G (2016) Territorial government reforms at the time of financial crisis: the dawn of metropolitan cities in Italy. Reg Fed Stud 26(4):509–530

MacLeod G (1997) Institutional thickness and industrial governance in Lowland Scotland. Area 29 (4):299–311

Magnaghi A (2010) Il progetto locale. Verso la coscienza di luogo. Bollati Boringhieri, Torino

Magnaghi A (ed) (2012) Il territorio bene comune. Firenze University Press, Firenze

Moulaert F, Mehmood A (2010) Analysing regional development and policy: a structural-realist approach. Reg Stud 44(1):103–118

Moulaert F, Nussbaumer J (2008) La logique sociale du développement territorial. Presses de l'Université du Québec, Québec

Moulaert F, Martinelli F, González S, Swyngedouw E (2007) Introduction: social innovation and governance in European cities. Urban development between path dependency and radical innovation. Eur Urban Reg Stud 14(3):195–209

Muro D (2009) Territorial accommodation, party politics, and statute reform in Spain. South Eur Soc Polit 14(4):453–468

Murphy AB (1989) Territorial policies in multiethnic states. Geogr Rev 79(4):410–421

Myrdal G (1957) Economic theory and under-developed regions. Duckworth, London

Nachmias D (1979) Public policy evaluation. St. Martin's Press, New York

O'Brien P (1982) European economic development: the contribution of the periphery. Econ Hist Rev 35(1):1–18

OECD (2013) Definition of functional urban areas (FUA) for the OECD metropolitan database, https://www.oecd.org/gov/regional-policy/Definition-ofFunctional-Urban-Areas-for-the-OECD-metropolitan database.pdf

Oppong JR, Ironside RG, Kennedy LW (1988) Perceived quality of life in a centre-periphery framework. Soc Indic Res 20(6):605–620

Palermo PC (2004) Trasformazioni e governance del territorio. Angeli, Milano

Pike AJ, Rodríguez-Pose A, Tomaney J (eds) (2010) A handbook of local and regional development. Routledge, London

Putnam R (1993) Making democracy work. Civic tradition in modern Italy. Princeton University Press, Princeton

Rivolin UJ (2005) Cohesion and subsidiarity: towards good territorial governance in Europe. Town Plan Rev 76(1):93–106

Rokkan S, Urwin DW (1983) Economy, territory, identity. Politics of West European Peripheries. Sage, London

Rostow WW (1960) The stages of economic growth. Cambridge University Press, Cambridge

Roxborough I (1979) Theories of underdevelopment. Macmillan, London

Sabatier PA (1999) The need for better theories. In: Sabatier PA (ed) Theories of the policy process. Westview Press, Boulder, pp 3–17

Storper M (1997) Territories, flows, and hierarchies in the global economy. In: Cox K (ed) Spaces of globalization. Reasserting the power of the local. The Guilford Press, London, pp 19–44

Trigilia C (2001) Social capital and local development. Eur J Soc Theory 4(4):427–442

Trigilia C (2005) Sviluppo locale. Un progetto per l'Italia. Laterza, Roma-Bari

Urwin DW (1991) Centro e periferia. Enciclopedia Treccani delle Scienze Sociali, http://www.treccani.it/enciclopedia/centro-e-periferia_%28Enciclopedia-delle-scienze-sociali%29/

Vaishar A (2006) Regional periphery: what does it mean? In: Komornichy T, Czapiewski K (eds) Europa XXI. Regional periphery in central and Eastern Europe. Warszawa, Stanislaw Leszczycki Institute of Geography and Spatial Organization, pp 7–12

Wallerstein I (1974) The modern world system. Academic Press, London

Williamson JG (1965) Regional inequality and the process of national development: a description of the patterns. Econ Dev Cult Chang 13(4):3–45

Zanon B (2014) Local development in fragile areas: re-territorialization processes in an Alpine community. Int Plan Stud 19(3–4):335–358

Chapter 2
Analysing the Quality of Territorial Policies in Europe's Periphery: The Theoretical and Methodological Framework

Abstract This chapter aims to develop the theoretical and methodological framework of the research. In the theoretical section of the chapter, we clarify the explanatory objectives of the study. First, we explain the nature of public policies and how they can be analysed. Second, we introduce the institutionalisation of the governance system as an independent variable that can provide descriptive and prescriptive explanations of the quality of territorial policies from both politics- and policy-oriented perspectives. We then present a unified policy-oriented theoretical framework, explaining the analytical dimensions of both the above-mentioned variables and the relationship between them. In the methodological section of the chapter, we first address the selection of the island of Sardinia as the territorial context of the research. Second, we present the multiple-case research design of the research and explain why urban regeneration and environmental protection were chosen as the policy areas to be addressed in the study. Finally, we discuss the choice of semi-structured interviews as the qualitative method of analysis to empirically test the research hypothesis.

Keywords Quality of public policies · Institutionalisation · Sardinia

2.1 Theoretical Issues

This book analyses the quality of territorial policies in a peripheral and disadvantaged area of the EU: the island of Sardinia. More specifically, our research focuses on why and how the quality of territorial policies may be improved in peripheral areas, exploring the factors and dynamics that improve the institutionalisation of the governance system of these policies. This chapter is organised as follows: First, we define some essential aspects of the dependent variable and specify what we mean by 'quality of public policies'. Second, we illustrate how political science scholars address the concept of institutionalisation as a major explanatory factor in the quality of public policies. Third, we define the institutionalisation of the governance system,

© Springer Nature Switzerland AG 2021

M. Tebaldi, M. Calaresu, *The Quality of Territorial Policies in Europe's Periphery*, Urban and Landscape Perspectives 22,
https://doi.org/10.1007/978-3-030-62647-1_2

providing a methodological explanation of why and how it can be used as the independent variable in three case studies of the quality of territorial policies.

2.1.1 Definition of the Quality of Public Policies and How It Can Be Analysed

In general terms, we can preliminarily define the quality of public policies as the set of features and properties that make possible the decision-making process to meet citizens' demands. Quality refers to the capacity, hence the probability, for obtaining a specific result that complies with specific standards or functional expectations of production processes and their related final output. Therefore, quality measurement responds to an evaluation logic, and it may refer to at least three different dimensions of public policies.

The first dimension is procedure-oriented, and it analyses the quality of decision-making and implementation processes. A quality public policy may cause political, social and institutional actors to feel responsible for ensuring (a) that laws are fairly enacted and (b) that there is a convergence between the original design of the policy and how it is implemented, starting from the decision-making processes and the inputs used for this purpose (resources, organisation, procedures, technologies, etc.). The aim of this dimension of quality assessment is to learn how, why, in what contexts and as a result of which factors public policies (a) adapt to environmental changes (i.e. policy *adaptability*); (b) acquire continuity over time without distortion, except incremental changes (i.e. policy *stability*); and (c) agree with each other harmoniously and consistently (i.e. policy *coherence*).

The second dimension is content-oriented, and it evaluates whether specific quality parameters of decision outputs and outcomes are achieved. An example of a content-related quality assessment is the set of information activities aimed at verifying the outcome (namely, the deterioration or improvement of services provided) of public policy (i.e. policy *performance*).

The third dimension is result-oriented, and it focuses on the extent to which a decision meets—and ultimately satisfies—citizens' demands. More specifically, a public policy is considered to be of high quality if it meets citizens' expectations for the performance of public decision-making. This means that it has to continuously (not only episodically) ensure both policy-makers' responsiveness to the needs and demands of society (i.e. policy *responsiveness*) and the welfare of the community rather than of individuals or small groups (i.e. *public regardness* of policy).

It is clear that the concept of quality in public policies, which is based on an evaluative effort, refers to theoretical models that focus contextually on two different pathways. The first is an explanatory analysis of those conditions that can produce or constrain the quality of decision-making and the mechanisms that can trigger it. The second is a prescriptive analysis of the behaviour and actions that allow such conditions to occur.

2.1.2 *Institutionalisation as an Independent Variable for Descriptive and Prescriptive Explanations of the Quality of Public Policies*

A comprehensive study of the quality of policy-making should combine explanatory and prescriptive analysis. In political science, one of the approaches that best suit both these goals is the neo-institutional approach. The scientific literature related to this approach agrees that (a) in explanatory–descriptive terms, the institutionalisation of public policies is relevant for explaining how the processes that affect the quality of the policies are activated (Gualini 2018) and (b) in prescriptive terms, institutionalisation is a factor that improves the quality of public policies and may be attained by configurable and enforceable actions by decision-makers (Huntington 1968; Polsby 1968; Ben-Dor 1975; Hult and Walcott 1990). Within the neo-institutional approach, descriptive and prescriptive capacities are shared by studies focused on democratic regimes, which draw a direct connection between the quality of public policies and the degree of institutionalisation of the actors, the rules and decision-making procedures. Therefore, according to the authors who share this approach, the primary explanation for the quality of public policies (a *dependent* variable) from both descriptive and prescriptive perspectives is, ceteris paribus, the institutionalisation of a series of attributes of the government system, or the governance system of policies (an *independent* variable), in a democratic regime.

The choice of the empirical referent of the independent variables (i.e. the government system or the governance system) depends on which approach researchers embrace to explain the quality of public policies. According to those who adopt a *politics-oriented* perspective, public policies are mainly conceived as the outputs of the government system. The government system itself can be conceptualised as the different ways the processes concerning the competition for political power are institutionalised. According to those authors, institutionalisation primarily concerns the organisational, procedural and regulatory configuration of democratic forms of government and their decision-making processes at different territorial levels.

Vice versa, according to those who adopt the *policy-oriented* perspective, the governance system of public policies is shaped by the different ways in which decision-making arenas are institutionalised. In this case, institutionalisation involves the organisational, procedural and regulatory characteristics of democratic governance and related decision-making processes both within and across different territorial levels.

2.1.3 *Politics-Oriented Perspectives*

Based on what is argued in the previous sections, it is worth asking what the requirements are for a democratic government system that can produce high-quality political decisions. This question is embedded in the classical tradition, especially Aristotle's political theory, in which he discusses the need to protect the 'government of the free' (called *πολιτεία*) from its possible degenerations, defined as the 'government of the multitude' (called *δημοκρατία* or *όχλοκρατία*). Aristotle resolved this issue with the fundamental principle that the *πόλις* political system must be guided by the rule of law, which must be above any citizen to prevent the subversion of the common good for partisan interests. In turn, the implementation of this principle depends on the system of institutions and rules governing the organisation of political power and collective decisions, the so-called constitutional government (or *πολιτεία*).[1]

Mutatis mutandis, the Aristotelian objective of preventing the abuse of demagogic power through the structures of the *πολιτεία* is pursued by those who, in contemporary political science, regard the institutionalisation of the forms and rules of the democratic governments as the main factor restricting the ineffective and inefficient practices of competition for and the management of political power to encourage those practices that are regarded as virtuous. Such authors base their analyses on the work of Samuel Huntington (1968, 1), who argues that 'the most important political distinction among countries concerns not their form of government but their degree of government'. In other words, Huntington emphasises that the comparison between political regimes, particularly in terms of their performance, must concern not only the *form* but also the *strength* of political institutions (i.e. the degree to which they maintain the order and stability of the political system). According to Huntington, the strength of the institutions depends on several factors, notably the degree of effectiveness with which institutions are able to regulate the behaviour of social actors, especially at the level of institutionalisation of organisations and procedures. Therefore, the level of institutionalisation is based in Huntington's view on four binary parameters: adaptability/rigidity, complexity/simplicity, coherence/disunity and autonomy/subordination.

The same logical pattern was applied by Eckstein (1971) in his pioneering analysis of the assessment of democratic performance. He measures the quality of public policies in terms of decision-making performance, that is, stability and public order. Eckstein operationalises those dependent variables using four indicators: the degree of government stability, the degree of decision-making effectiveness, the level of violence, and the degree of institutional legitimacy. They all refer to the conforming action of a political–institutional *explanans* (government institutionalisation) while keeping in mind the impact of others' social economic and cultural intervention variables.

Jackman (1993) uses a similar approach in his comparative study of political development, which is intended primarily to examine the decision-making efficiency

[1] According to Aristotle, the constitution (*πολιτεία*) concerns the magistrates (*αρχαία*), the way they are assigned, the assignment of sovereignty (*τὸ κύριον*) and the determination of the purpose of each association (*κοινωνία*). It is separate from the laws (*νόμοις*) laying down the rules under which the judges exercise their power (*ἄρχειν*) and punish offenders.

of national governments. Such decision-making performance is expressed as the product of the government's political capacity. The concept of political capacity is divided into two different analytical dimensions. The first dimension is the organisational persistence of government institutions (the higher the stage of development of the decision-making institutions in terms of their organisational age, the greater their adaptability and, therefore, their ability to produce decisions continuously). The second dimension is legitimacy, which is conceived as the degree of compliance and consensus that government institutions and political leaders are able to obtain. As Jackman argues (1993, 95):

> Legitimacy is fundamental to political life because it reflects the degree to which those who seek to rule (i.e., to exercise power) are accepted by the ruled. It thus identifies the nature of the relationship between rulers and ruled.

Przeworski (1995) embraces a similar approach. He states that the consolidation of high-quality democracies, which are more likely to survive adverse environmental conditions, depends on how the institutionalisation of government systems gives them decision-making capacity. Such governance capacity results in high performance to control arbitrary violence, resolve social conflict or improve social and economic conditions. Notably, in addition to sharing a logical framework and key hypotheses, all the authors mentioned above make an effort to explain the effects (e.g. the quality of some key policies that synthesise the performance of the entire democratic system) by connecting them to the same intermediate cause (the capacity or strength of the democratic governing institutions), which depends on the degree of institutionalisation of the government system. Below is a simplified explanation:

degree of institutionalisation of the government system → capacity or strength of governing institutions → quality of public policies

Among the scholars who recently followed a politics-oriented perspective, it is worth mentioning the work conducted by the Latin American and Caribbean Research Network within the Research Department of the Inter-American Development Bank. The theoretical framework adopted by these authors (Scartascini and Tommasi 2012; Tommasi and Scartascini 2012; Scartascini, Stein and Tommasi 2013;) can be summarised as follows:

degree of institutionalisation of the government system → tendency for long-term cooperation among actors → quality of public policies

According to this framework, the quality of public policy consists of a set of features (responsiveness, stability, adaptability, coordination and coherence, implementation, efficiency, orientation towards the public good) that can be realised only in contexts where long-term cooperation among the involved actors is possible. Such cooperation is more likely to occur if the government systems where political exchanges take place are highly institutionalised.

2.1.4 Policy-Oriented Perspectives

While based on solid theoretical foundations, the politics-oriented perspectives only partially explain what quality of public policies means in democratic regimes and where it should be sought (see Tebaldi and Calaresu 2009, 2015, 2016, for a review). It also raises some questions, which are well highlighted by Plattner (2005). In particular, Plattner notes that on the one hand, the quality of democratic decision-making refers to mechanisms and results that are typical of a *vertical democracy* based on the fundamental relationship between the representatives and the represented in a framework of politics-oriented relations; on the other hand, the quality of democratic decision-making leads us to consider the effects of *horizontal democracy*, which includes the arenas and products of governance in a policy-oriented relationship framework. The influence of other relational dynamics may also be highlighted in this dimension because, in several policy areas, the decision-makers are often not affected by the rewards and sanctions allocated in the electoral representative setting. Therefore, it seems reasonable to explore the fields of study that share institutionalisation as an independent variable but address the quality of public policy from different perspectives. They all focus on the arenas and processes of policy-making in order to produce empirical evidence about the influence of institutionalisation of the governance system on the production of high-quality policies.

Political scientists have long openly questioned and rejected both the descriptive and prescriptive values of the harmonious dichotomy between politics and public administration inherited from juridical doctrines and democratic structural functionalist theory.[2] This analytical perspective stems from the following theoretical connection between Schumpeter and Weber, whom both distinguish different *moments* and *times* of democracy. On the one hand, the theory of competitive leadership (Schumpeter 1943) tends to focus on the input of the democratic process: the moment of electoral competition and, closely related to its results, the legitimacy and the selection of the rulers. However, Schumpeter's theory does not tackle the determination of the content and modes of government. On the other hand, the Weberian theory of legal–rational authority, the description of the output of the democratic process, which enables the daily management of government decisions between elections, is entrusted to the hierarchical–linear model of government. Elected officials are responsible for making political decisions and producing public policies, whereas bureaucrats are responsible for the implementation of those decisions.[3]

[2]In this context, Hjern and Hull (1982) speak of methodological constitutionalism or conventional wisdom. This has led individuals in the fields of political science and public administration to often interpret regulatory architecture and constitutional powers not only as a prescriptive element of the expected social order but also (mostly) as a descriptive tool of interactions between democratic institutions.

[3]Monocentric policy-making concepts, like those of party government (Rose 1974; Castles and Wildenmann 1986; Blondel and Cotta 1996, 2000) and bureaucratic government (Page 1985; Peters 1981), were developed based on this mainstream theory.

The weak descriptive capability of this theoretical model requires us to extend the analytical perspective to all actors who, although they lack electoral legitimacy and are not strictly defined as governmental actors, tend to shape and influence policy-making processes, especially to define fundamental aspects of the quality of democratic policies. Due to this explanatory gap, public decision-making processes that are able to go beyond the monocentric perspective of a *party government* or a *bureaucratic government* should be used to develop pluralistic images and schemes stemming from the concept of governance.[4] However, it is worth questioning how the design of governance can improve the quality of public policies. An answer is provided by authors who, while focusing on explanatory goals, highlight which conditions and organisational pathways could lead to a governance model that offers good decision-making performance and high levels of legitimacy while minimising the potential causes of democratic failure (e.g. exclusion of unorganised citizens, increased social inequality, weakening of public authority, exacerbation of distributive collusion).

One set of studies focuses on the long-term conditions capable of shaping the inter-organisational configuration (i.e. policy network) of governance. According to some of these authors (Thelen 2003; Pierson 2004), the outcomes of the decision-making performance of governance depend on those conditions and, in particular, on their institutionalisation. This logical scheme can be summarised as follows:

institutionalisation of the policy network → *quality of public policies*

The institutionalisation process of policy networks is rooted in the historical, economic, political and social conditions in which they operate. To properly understand the current morphology of the organisational fields of public policy, it is first necessary to trace the historical factors that characterised their establishment and consolidation. A scientific explanation of the current forms of governance must take into account their evolution, especially the evolution of their inter-organisational dynamics, starting from an *original model* of interactions. With appropriate adjustments, we may support scholars who suggest that inter-organisational relations and the organisational characteristics of public policies depend on the history of how the organisational field has been generated and consolidated. How a policy network emerges, or the *pillars of its creation*, can influence its inter-organisational

[4]In the governance model, collective decisions result from negotiations and interactions between decentralised agreements among public institutions at different levels and the actual powers that control coalition, financial, professional, informational and symbolic resources. Private organisations and public authorities are connected by continuous negotiation processes that redefine the authority derived from voting, resulting in a variable balance between unilateral command acts and multilateral voluntary agreements. There is an extensive literature on the concept of governance, ranging from political science to public administration to organisational sociology. However, the purpose of our work is not to shed light on the different analytical dimensions underlying the use of this concept in the social sciences. For a general discussion of the field, see the work of Kooiman (1993), Rhodes (1996, 1997), Pierre (2000) and Peters and Savoie (2001).

Table 2.1 Characteristics of the three pillars of institutions

	Regulative	Normative	Cultural-cognitive
Basis of compliance	Expedience	Social obligation	Taken-for-grantedness
			Share understanding
Basis of order	Regulative rules	Binding expectation	Constitutive schema
Mechanisms (of control)	Coercive	Normative	Mimetic
Logic	Instrumentality	Appropriateness	Orthodoxy
Indicators	Rules	Certification Accreditation	Common beliefs
	Laws		Shared logics of action
	Sanctions		Isomorphism
Affect	Fear Guilt/ Innocence	Shame/Honor	Certainty/Confusion
Basis of legitimacy	Legally sanctioned	Morally governed	Comprehensible Recognisable
			Culturally supported

Source: Adapted from Scott (1995)

attributes even decades later.[5] The assumption that each inter-organisation network is influenced by the political and administrative decisions of its founders applies to all fields of policy (Thelen 2003). In other words, policy networks are based on the decisions that have moulded their original form. The original model (i.e. the set of structural and behavioural features that are configured when the policy network was founded and that underpin the process of institutionalisation) is a crucial part of the life cycle of an organisation. Empirical evidence shows that the original model tends to reproduce itself over time and affects inter-organisational relations even after several cycles of organisational field development (Pierson 2004).

In addition to the institutional core actors, administrative structures and their formal relations, diachronic–longitudinal analysis of governance seeks to uncover the changing context within which political and institutional actors interact. Specifically, it investigates the cognitive, regulative and normative components—the three pillars of institutions identified by Scott (1995)—those guide decision-makers who must choose between technically complex and ideologically controversial alternatives (see Table 2.1).

The three pillars of institutions ultimately result from the interactions among political and administrative bodies and economic, cultural and interest organisations that participate in the formation of authoritative decisions. These exchange relationships prompted the development of currently used and socially accepted knowledge

[5]See the seminal studies of Granovetter (1985) and Powell and Di Maggio (1991).

Table 2.2 Type of evaluation processes

		Links between the state and social actors in the policy subsystem	
		Several	*Few*
Administrative capacity of the state	*High*	Social learning	Lesson-drawing
	Low	Formal evaluations	Informal evaluations

Source: Adapted from Howlett and Ramesh (1995)

(*cognitive* component), institutional projects (*regulative* component) and values and criteria of legitimacy (*normative* component), to which policy-makers usually refer during the public policy cycle.

Another set of studies, inspired by Lindblom (1990) and Lindblom and Cohen (1979), explores the causal relationship between the institutionalisation of the public policy evaluation process and the quality of policies (Lopez-Acevedo, Krause and Mackay 2012) through the following scheme:

institutionalisation of public policy evaluation process → *political–institutional learning and adjustment capacity* → *quality of public policies*

The authors who investigate this field of study share a basic assumption: if disseminated and accessible information generates constant improvement of decision-making processes, then evaluation plays a central role because it is a primary, and equally constant, source of learning and information. Evaluation is a form of knowledge that can be used to support social interdependence and, therefore, is a primary source of democratic and good quality public decisions. In other words, results and evaluation processes become important to the quality of public policies when they serve as institutionalised knowledge bases for the actions taken during all stages of political and administrative activity and, possibly, when major adjustments may be required based on the data that are examined, learned and evaluated.

Therefore, the institutionalisation of evaluation processes and policy change are intimately linked. However, in order to understand how and to what extent institutionalised evaluation processes may influence policy change and therefore achieve better quality decision-making, it is necessary to investigate the relationship between evaluation processes and institutional–political learning.[6] This is because learning filters the information obtained and steers the process of change (Bennett and Howlett 1992; Feick 1992; Leeuw, Rist and Sonnischen 1994). In this regard, Howlett and Ramesh (1995) adapted the study of Cohen and Levinthal (1990) and developed a matrix (Table 2.2) to highlight how and why evaluation processes could be a more or less powerful vehicle for learning and thus affect the propensity for policy change.

[6]Learning has been analysed in the fields of social and political science from different perspectives. For some of the most significant contributions from these perspectives, see the work of Sabatier and Jenkins-Smith (1993), Hall (1993) and Huber (1991).

According to Howlett and Ramesh (1995), there are two aspects of institutionalisation that explain the propensity for learning and the probability that the learning will be triggered by policy evaluation processes. The first is the administrative capacity of the state, and the second is the morphology of the policy network. The first variable concerns the administrative structure, including the entire set of valuable resources (e.g. financial, legal, technical and, above all, cognitive–symbolic resources) that state bureaucracies can mobilise in policy-making processes. The other variable concerns the density and strength of the ties among political–bureaucratic actors and social actors that can be structured in each policy network. For the evaluation to trigger learning mechanisms and (through them) steer political and social change processes characterised by diffused knowledge and specialised competencies, the administrative capacities of the state must be high and highly institutionalised. Only then can assessment tools generate usable knowledge that enables institutional learning or social learning.

As far as the quality of public policies is concerned, the formal and informal evaluations implemented by administrative actors that are weakly institutionalised and have poor information absorption capacity seem irrelevant. In particular, when a high capacity for public administration emerges within state-centric, highly institutionalised policy networks that are able to regulate the access of other social actors to decision-making arenas, then either endogenous or lesson-drawing institutional learning occurs (Rose 1991, 1993). In these cases, learning is internal to political-administrative agencies' policy processes within small and technically skilled policy networks. By virtue of their strength, governmental institutions transform evaluation into learning. In turn, the latter becomes the central mechanism to adjust decisions that are made or fully change them, if necessary, in light of the objectives that generated them (Hall 1993).

2.2 Looking for a Unified Policy-Oriented Theoretical Framework

As shown above, within the context of policy-oriented analysis, several studies correlate (in an unsystematic and fragmentary way) the quality of public policies to the institutionalisation of the inter-organisational relationships and patterns between the state and social actors. Our theoretical effort aims to take a step towards creating a systematic single analytical framework based on the theoretical fragments that emerged from the review of the policy-oriented literature presented in the previous section. More specifically, our theoretical proposal aims to explain the quality of public policies as a *dependent* variable of the degree of institutionalisation of their governance system, which is an *independent* variable.

2.2.1 The Relations Among the Institutionalisation of the Governance System and the Quality of Territorial Policies

The governance system of a public policy refers to a decisional configuration that consists of organisational actors linked by ties of varying intensities and is oriented to systematically and continuously produce (through the development of such ties) binding decisions on public issues in a specific policy field.

A governance system can be examined in terms of institutionalisation by taking into consideration its organisational configuration. This dimension relates to the inter-organisational structure of the governance system, and its main empirical reference is the composition of the dominant coalition and the gatekeeping barriers of the policy field.

A *dominant coalition* includes all the actors who are able to achieve consensus and legitimacy for their leadership and the content of their decisions. It may involve potentially conflicting interests and heterogeneous logics of action, creating a positive-sum game for the other actors in the organisational field by offering programmes and alternative choices. Although this is likely to happen, it is not certain that the pivotal role within the dominant coalition will be performed by one or more institutional actors. Therefore, it is possible that the task of mobilising consensus and shaping the organisational field would be carried out by non-institutional actors endowed with specific technical or knowledge resources (e.g. experts, opinion leaders, members of epistemic communities) or substantial financial resources (e.g. industrial groups, corporations). Therefore, the dominant coalition's legitimacy essentially depends on the availability, control, acknowledgement and capacity to use key resources. The relevance of these resources is connected to (a) the type and homogeneity of the setting in which most of the exchanges between actors in the organisational field occur (e.g. ideological–regulatory, technological–rational, procedural–bureaucratic settings) and (b) the capacity of the dominant coalition to act cohesively and to rule over the other actors in the policy network effectively.

Gatekeeping refers to the barriers that restrict access to policy networks. More specifically, it constrains the capacity of actors outside the dominant coalition to influence the outcomes of decision-making processes. From an empirical perspective, gatekeeping can be intended as the set of rules and technical, economic and organisational standards that form the threshold to be overcome to have an influence over decisions.

The degree of gatekeeping is an indicator of the level of impermeability of the dominant coalition compared to the organisational field. Regarding this analytical dimension (the inter-organisational structure of the governance system), we hypothesise the following:

H0: The greater the (a) functional specialisation and clarification of maps of interdependence, (b) congruence of cognitive patterns within the dominant coalition (particularly between those who decide and those who implement the decisions) and (c) structuring and impermeability of gatekeeping, the higher the level of institutionalisation of the governance system of a policy.

Furthermore, we can assume a direct correlation between the quality of a particular type of policy (as a dependent variable) and the degree of institutionalisation of its governance system (as an independent variable). The degree of institutionalisation of the governance system can influence three fundamental analytical dimensions of the quality of public policies. Thus, we hypothesise the following:

H1: The degree of institutionalisation of the governance positively affects the responsiveness of a policy to social issues.

As Mathur and Skelcher (2004) suggest, highly institutionalised contexts have an ability to filter, interpret, tune and constantly adjust stakeholders' requests, according to the response or non-response time. This is a result of the development of tools and regulatory standards (i.e. constitutional hardware)[7] and of the control of cognitive maps and epistemic resources (i.e. constitutional software)[8] by actors who interact within the governance system. Concerted policies, which govern broad categories of interests, are frequent, and they are usually expressed over the medium to long term with macroeconomic effects. On the contrary, public policies are likely to have undefined aims and goals if the degree of institutionalisation of the governance system is low. In areas with low institutionalisation levels, actors outside the dominant coalition may cause frequent and sudden changes in the set of resources and their stakes and, therefore, in the balance of power among actors within the policy network. The vulnerability of dominant coalitions and a low degree of gatekeeping tend to generate fragmented, disorganised policies, and the decision-making processes and their implementation are affected by specific constellations of groups seeking an appropriation of disaggregated resources (Foweraker and Landman 2002). This leads us to the following hypothesis:

H2: The degree of institutionalisation of the governance system positively affects policy network actors' propensity for cooperation.

As the degree of institutionalisation of the governance system increases, public policies tend to experience a *depoliticisation* process. As a consequence, many of the highly ideological and political issues of the governance system become routinised issues, and decisions are made at the administrative level. A policy generated by a high degree of institutionalisation of the governance system also reduces the degree of conflict among actors in the policy network. This is because many ideological and political issues are no longer the subject of constant renegotiation among actors in

---

[7]See De Bruijn (2002); Ingraham, Joyce and Donahue (2003); and Foweraker and Landman (2002).

[8]See Yanow (1996; 2000) and Dryzek and Berejikian (1993).

the policy network since the general lines of action have already been clearly defined and accepted.

If the degree of institutionalisation is high, then the allocative effects of the decisions taken typically do not generate opposition from non-institutional actors given the high degree of legitimacy of the dominant coalition. In particular, a high degree of institutionalisation is able to generate high levels of loyalty since potential challengers of the dominant coalition fit within a system of structural relationships that is rooted in procedural and normative ideological constraints (Ingraham, Joyce and Donahue 2003; Koppenjan and Klijn 2004). This occurs when those systems of structural relationships produce collective identities and forms of membership that intertwine with policy sub-cultures and generate an orientation towards the common interest (Scharpf 1989). On the contrary, low institutionalisation does not allow for generalised compliance of social groups with decisions made by the dominant coalition. Hence, it frequently results in political and social protests, boycotts or non-compliance with the rules. Based on this, we hypothesise the following:

H3: The degree of institutionalisation of the governance system positively affects the degree of functional congruence of policies.

When the degree of institutionalisation of the governance system is high, the resistances and filters imposed by the dominant coalition to hinder exogenous pressures by small groups are strengthened by both the ability to establish stable lines of action—which are taken for granted—and the high probability of transforming the social or institutional context (i.e. the target of public policy) without having to constantly renegotiate the terms of agreements that are reached. First, this allows policy-makers to define long-term programmes, which are mostly consistent and not reversible over time. It also helps restrict territorial or regional imbalances, which are frequently generated by transition from the decision stage to the implementation stage because of the different territorial levels that interact in the two stages of policy-making (Fischer 2000; Rosenau 2000; Lynn, Heinrich and Hill 2001). Vice versa, when the degree of institutionalisation of the governance system is low, then the degree of functional congruence of policy is also low, depending on the resources of strong economic stakeholders, the organisational capability of small economic groups, the conflict/alliance relations established among such actors and the type of relations that bind them to the bureaucracy. The result of a low degree of institutionalisation is pervasive particularism that leads to different models of policy implementation at the local level (Matland 1995), to a low degree of integration among programmes and subprograms and to the presence of contradictory elements in different normative and administrative measures.

In a nutshell, a high degree of institutionalisation of the governance system would likely result in a higher-quality public policy. This is because of the high degree of responsiveness, tendency to build cooperative interactions among actors and proper, functional consistency between planned actions and implementations. By contrast, a low degree of institutionalisation of the governance system would likely result in a lower-quality public policy due to the limited capability of the dominant coalition to effectively respond to social problems, the tendency for conflicting interactions among actors in the policy field and poor functional consistency between planned actions and implementations.

2.2.2 *The Unified Policy-Oriented Theoretical Framework*

We propose here a unified, policy-oriented theoretical framework that tries to recompose an established but fragmented theoretical framework in policy studies, as illustrated in the previous section. The main assumption of our framework is that the institutionalisation of the governance system is an independent variable that is capable of influencing the quality of public policy, intended as a dependent variable. According to the framework, the two above-mentioned variables tend to change according to a linear and direct correlation: upward variation of the degree of institutionalisation of the governance system tends to correspond to upward variation in the quality of the policy, and vice versa.

In order to empirically analyse the degree of institutionalisation of the governance system, we split up the concept into three analytical dimensions: (a) *functional specialisation and interdependence maps among the actors of the dominant coalition*, (b) *congruence of cognitive maps within the dominant coalition* and (c) *structuring and impermeability of gatekeeping*.

The higher the degree of these three dimensions, the higher the level of institutionalisation of the governance system of the policy. Each dimension is assessed according to a five-level scale (i.e. low, medium/low, medium, medium/high, high).

We also split up the concept of quality of public policy into three more analytical dimensions: (a) *responsiveness*, (b) *cooperation* and (c) *functional congruence*. The higher the degree of these three dimensions, the higher the quality of a public policy. Each dimension is assessed according to the same five-level scale.

As a final step, we conduct an overall assessment of the two variables based on the average values of the three analytical dimensions for each concept on the five-level scale. The unified policy-oriented theoretical framework is outlined in the following graph (Fig. 2.1).

2.3 Methodological Issues

This section considers the main methodological issues addressed by our research and the methodological solutions we adopted. First, we analyse the territorial context under consideration (the island of Sardinia), highlighting its demographic and social–economic characteristics in order to support the choice to focus on Sardinia. Second, we explain the logical framework that led us to focus on two specific areas of policy that are particularly relevant in the context of territorial policies: urban

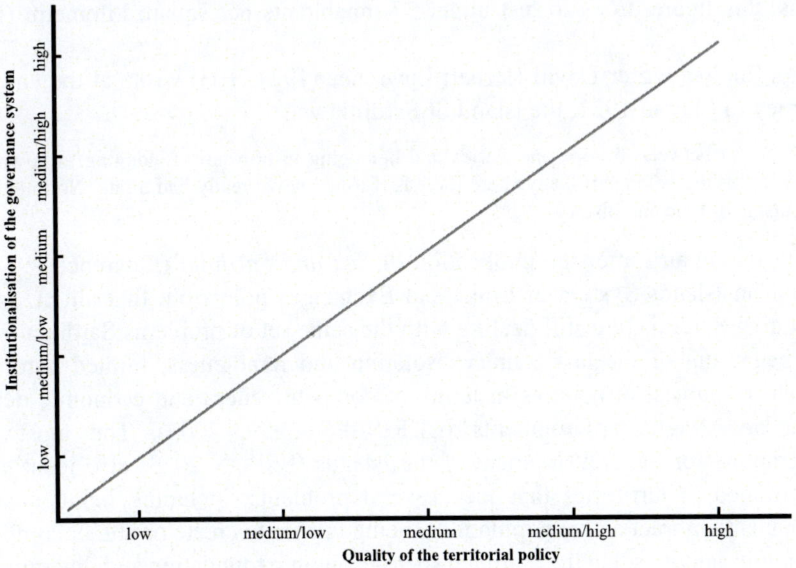

Fig. 2.1 Unified policy-oriented theoretical framework

regeneration and environmental protection policies. Third, we explain how we selected three typical cases.

2.3.1 *The Island of Sardinia as the Territorial Context of the Research: A Demographic and Socio-economic Profile*

The island of Sardinia is an administrative territory with a special regional status (*Regione Autonoma della Sardegna/Regione Autònoma de Sardigna*). With a surface of 24,100 square kilometres, Sardinia is the second-largest island in the Mediterranean Sea after Sicily, and it is the second most populated (ISTAT 2012). More than 40% of its 1,661,723 inhabitants, who are distributed in 8 provinces and 377 municipalities (ISTAT 2014), live in two FUAs of local importance: one on the north side of the island (the FUA of Sassari) and one on the south side (the FUA of Cagliari). The geographical centre is experiencing progressive depopulation (Bottazzi, Puggioni and Zedda 2006). With 68.95 inhabitants per square kilometre, Sardinia also has the lowest overall regional population density in Italy and one of the lowest densities among the Mediterranean islands (Eurostat 2011). In inland areas, this figure drops to just under 37 inhabitants per square kilometre (ISTAT 2014).

As English writer David Herbert Lawrence (1921, 103) wrote at the end of his journey to Italy in 1921, the island of Sardinia was:

> [...] lost between Europe and Africa and belonging to nowhere. Belonging to nowhere, never having belonged to anywhere [...] as if it had never really had a fate. No fate. Left outside of time and history.

Almost a century after the publication of *Sea and Sardinia* (Lawrence 1921), the European Islands System of Links and Exchanges points out that all EU islands, including Sardinia, are still dealing with the same set of problems Sardinia faced at the beginning of the last century: isolation and remoteness, limited natural and human resources, difficulties in terms of competitiveness and economic development and fragile environments (EURISLES 1999, 2003). The most recent Programme for the Development of the Islands (ESPON 2013) also identifies the EU islands as territories that face several problems, including being unable to permanently attract active populations, being unable to create or attract competitive economic activities, having a limited socio-economic foundation and diverging from the EU's objectives of sustainable development and economic, social and territorial cohesion.

More than a third of the Sardinian population lives in economically depressed areas, and the island as a whole has many of the backward socio-economic characteristics observed in southern Italy (Leonardi 1995; Barca 2001). In 2013, Sardinia's unemployment rate was 17.5%, 2% increase from 2012 (CreNos 2014). Compared to the European average, which stood at about 7%, Sardinia's unemployment rate looks dismal. According to the most recent studies, Sardinia's record is weak even in comparison to other peripheral and crisis areas; the average unemployment rate of other EU islands is 11.6%.

The Sardinian economy's relative standing in the EU continues to decline as the number of indicators associated with the more peripheral areas of Europe increases. In terms of wealth produced, Sardinia is 190th out of 272 European regions, with a regional GDP per capita just 77% (2014) of the European average, dropping from 78.3% in 2010 and 80.3% in 2009 (CreNos 2014). According to the 2nd EU Report on Regional Competitiveness (Annoni and Dijkstra 2013), which considers a diverse set of factors in 262 regions of 28 European countries, Sardinia ranks 222nd. In terms of the disaggregated data, the island's performance appears to be negative, especially with regard to institutional factors. In a list of 259 regions, Sardinia never ranks above 219th. Regarding aspects related to infrastructure, particularly road and rail accessibility, the index ranks Sardinia as 231st out of 259 European regions.

To diminish its demographic, socio-economic and environmental issues, the island of Sardinia would need to achieve 'good enough governance' (Grindle 2004, p. 526). At the same time, however, several scholars underline that peripheral areas like Sardinia experience the most difficulties in achieving good governance. Many EU peripheral areas simply do not have the (pre)conditions for institutional learning, innovation and growth (Hospers 2003) due to structural and institutional problems (Rodríguez-Pose 1998, 1999; Calaresu, De Micheli and Tebaldi 2018).

Of the islands in the EU, the island of Sardinia was selected as a case study for analysis for three main reasons. First, research results obtained from this island are expected to have more representativeness, significance and generalisability in comparison to other peripheral EU areas (particularly other islands). Second, it has specific relevance in the European setting. The available socio-economic data suggest that although, as already mentioned, Sardinia is considered to be a fully peripheral area (see below) that deserve to be studied due to its particular characteristics and problems. Third, despite its relevance and specificity, Sardinia has been studied little up to date (Acciaro 2008), as if it were still 'without a fate' and 'out of time and history'.

2.3.2 *The Multiple-Case Research Design*

To examine the territorial context of the island of Sardinia, we adopt a *multiple-case framework*, in which two or more cases are selected within the same research, and each case can be considered a *replica* of the other one. If properly selected, a group of cases is a significant analytical foundation upon which to empirically test hypotheses and theories. For this reason, in a multiple-case research design, the choice of each case must follow a careful sampling logic. Our sampling logic is based on the principle that the selected units (the *sample cases*) are intended to represent a wider universe. The selection of cases is, therefore carried out in accordance with representative criteria previously identified with a solid theoretical basis and classification. To narrow down the selection of our case studies, we *drew* concentric circles. In the largest circle, we made an initial choice concerning the types of territorial policies to be examined. The middle circle determined which policy areas to be analysed for each type of territorial policy. The narrowest circle identified the case study to focus on within each policy area.

2.3.2.1 Planning, Regulation of Use and Transformation of the Territory as Territorial Policy

As mentioned in the first chapter (Chap. 1), this study focuses on one type of territorial policy that involves the planning, regulation of use and transformation of the territory. The reasons for this choice stem from the socio-economic characteristics of the territorial context we analyse. Sardinia is an island affected by several economic and social issues. Among European peripheral areas, it ranks in the lowest positions, especially for economic and infrastructural development indicators. Regarding a possible solution to these issues and identifying possible obstacles, our research starts by analysing the most appropriate territorial policies in terms of representativeness and significance, that is to say, policies that are specifically oriented to solve the aforementioned issues through public actions and programmes. In terms of territorial policies, we choose to examine those that establish inclusion criteria to use public goods and reduce disparities arising from unequal use of the territory. These policies are most able to influence territorial planning, use and transformation (e.g. urban transformation and environmental protection policies).

2.3.2.2 Urban Regeneration and Environmental Protection as Policy Areas

For each type of territorial policy, we identify policy areas that serve as subjects for the empirical analysis in this research. Within the context of territorial planning, use and transformation policies, we examine urban regeneration and environmental protection policies.

These two policy areas are significant for the territory of Sardinia due to their threefold strategic profile, which is (a) economic–functional, (b) social and (c) political.

From the economic–functional point of view, both sectors must be considered crucial to the economy of the island, especially its social and economic development. In addition to affecting the tourism economy, environmental protection policies have a strong impact on agriculture, livestock farming, maritime activities and fisheries, infrastructure and housing. Urban regeneration policies also have significant effects on the construction and hospitality industries.

From the strategic social perspective, environmental protection policies undoubtedly have social strategic importance. This is especially true when considering the benefits of such policies for the well-being of citizens, particularly those in less urbanised rural and coastal areas. Urban regeneration policies are particularly important for their potential social impact on the overall wealth of populations living in urban areas.

From the strategic political point of view, protection of the most valuable natural landscape areas, such as coastal and mountain areas, has been a goal of national, regional and local governments for several decades as well as the mission of many groups and collective movements. Urban regeneration is one policy area in which the public debate on planning, spending and regulation of local and regional governments has been more lively in the last decade than ever. In particular, this debate has focused on the promotion of Old Towns, the locationing of construction areas for residential and industrial use and restrictions on building volume.

For these strategic reasons, the two policy areas are selected in response to the methodological criterion illustrated above. In addition to restricting the analysis to two policy areas, we must select which decision-making processes deserve to be studied within each area. As mentioned in the methodological section, our guiding criterion for this selection is the representativity of the case study for each policy area.

2.3.2.3 Three Typical Cases as Case Studies

As Seawright and Gerring argue (2008, 294), 'case selection is the primordial task of the case study researcher, for in choosing cases, one also sets out an agenda for studying those cases'. The issue of selecting a case study arises when the selected case aims to represent, with its specific characteristics, a larger population of cases (Gerring 2004, 2007).

Of the various methods to select a case study to represent each policy area, we adopt the *typical case method* (Eckstein 1975). This method bases the choice of a case study on the fact that the selected sample is a comprehensive example of all the possible variations in the population of the cases that it represents. In other words, the selected case study is considered an exemplary case that is capable of representing all other cases that fall into the same class of phenomena, as it includes all the relevant dimensions of analysis that are necessary to study that class. The advantage of using this case selection method is identified by Seawright and Gerring (2008, 299) as follows:

> Because the typical case is well explained by an existing model, the puzzle of interest to the researcher lies *within* that case. Specifically, the researcher wants to find a typical case of some phenomenon so that he or she can better explore the causal mechanism at work in a general, cross-case relationship.

We conduct preliminary interviews with a panel of experts to identify typical case studies that can well represent the transformations within the two policy areas that took place in the period between the early to mid-1990s and the first decade of the 2000s. Based on the experts' answers, we select typical cases for each area.

Regarding the environmental protection policy area, we analyse marine reserve protection as part of the broader government agenda to protect marine environments by defining marine protected areas (MPAs hereafter). The typical case we analyse is the Sinis peninsula – Mal di Ventre island MPA, which was established in the Municipality of Cabras in 1997. Our analysis covers the period from 1993 to 2004.

Typical cases of urban regeneration processes in the Old Towns of the municipalities of Sassari and Alghero are also subjected to empirical analysis. For both cases, the analysis covers the period from 1995 to 2006.

2.3.3 *Semi-structured Interviews as a Qualitative Method of Analysis*

A multiple-case research design lends itself to in-depth analysis of innovative theories and hypothesis, especially when preliminary empirical tests are run to test the robustness of the logical links that bind the independent variable to the dependent variable. To produce the best knowledge effects, in-depth analysis tends to be associated with qualitative methods, such as semi-structured interviews. For the purposes of this study, semi-structured interviews may be the most suitable instrument since they allow for a high degree of control over the context in which the research study is performed. This is due to the continuous presence of the researcher–interviewer *in the situation*, similar to what happens in participant observations. This enriches our research study with information that is otherwise inaccessible through the use of *cold tools*, such as surveys, which greatly diminish the connection between the object of the research and the researcher.

For the purposes of this work, we use a qualitative tool with an intermediate degree of structure. This tool enables us to overcome both the lack of comparative reliability of unstructured interviews and the insufficient information depth of rigidly standardised interviews. While planning the interview outline to be submitted to different categories of selected actors, we chose the so-called interview guides. Following Goode and Hatt (1952), we use this term to refer to flexible patterns of pre-set questions posed in a particular sequence. These schemes aim to collect the same information from each respondent, but they allow the interviewer to repeat the question in a flexible manner, changing words in accordance with the statements given during the interview and the type of interaction that develops with the interviewee. The instrument's flexibility gives the interviewer both the ability to express questions in a way that allows the interviewee to understand them as easily as possible and the ability to ask deeper questions if needed.

In this research, we apply the qualitative method defined as the 'snowball technique' (Moyser 2006), through which it is possible to create a panel of key actors who are involved in the decision-making processes under analysis. The snowball technique combines both positional and reputational selection methods. The key actors to be interviewed are initially selected according to a criterion that takes into account the roles of power at the local level, where the public policies were first developed and then implemented. We select each interviewee on the amount of social or political/institutional power they held during the examined timeframe. Then, they serve as judges who nominated additional individuals to be interviewed based on their reputation. We choose to conduct the interviews in person with face-to-face interaction between the interviewer and interviewee. When this is not possible, the interviews are conducted via mobile phone or with digital tools, such as Skype. Information are collected and recorded via magnetic audio support and then transcribed to electronic media and cross-checked (see Appendix for further information).

References

Acciaro M (2008) The role of ports in the development of Mediterranean islands: the case of Sardinia. Int J Transp Econ 35(3):295–323

Annoni P, Dijkstra L (2013) 2nd EU report on regional competitiveness. European Commission Joint Research Centre, Luxembourg, http://www.jrc.ec.europa.eu/

Barca F (2001) New trends and policy shift in the Italian Mezzogiorno. Daedalus 130(2):93–113

Ben-Dor G (1975) Institutionalisation and political development: a conceptual and theoretical analysis. Comp Stud Soc Hist 17(3):309–325

Bennett C, Howlett M (1992) The lessons of learning: reconciling theories of policy learning and policy change. Policy Sci 25(3):275–294

Blondel J, Cotta M (eds) (1996) Party and government. Macmillan, London

Blondel J, Cotta M (eds) (2000) The nature of party government. Palgrave, London

Bottazzi G, Puggioni G, Zedda M (2006) Dinamiche e tendenze dello spopolamento in Sardegna. Centro Regionale di Programmazione-Regione Autonoma della Sardegna (CRP), Cagliari

Calaresu M, De Micheli C, Tebaldi M (2018) Institutionalization of the governance system and the quality of public policies. A case study on public transit and local transportation. Riv Ital Politiche Pubbliche 13(1):91–126

Castles FG, Wildenmann R (eds) (1986) The future of party government. 1: visions and realities of party government. De Gruyter, Berlin/New York

Cohen WM, Levinthal DA (1990) Absorptive capacity: a new perspective on learning and innovation. Adm Sci Q 35(1):128–152

CreNos (2014) Economia della Sardegna, XIV Rapporto 2007. CUEC, Cagliari

De Bruijn H (2002) Performance Management in the Public Sector. Routledge, London

Dryzek JS, Berejikian J (1993) Reconstructive democratic theory. Am Polit Sci Rev 87(1):48–60

Eckstein H (1971) The evaluation of political performance. Sage, London

Eckstein H (1975) Case studies and theory in political science. In: Greenstein FI, Polsby NW (eds) Handbook of political science. Vol. 7. Political science: scope and theory. Addison Wesley, Reading, pp 94–137

ESPON (2013) Programme for the development of the islands – European islands and cohesion policy (EUROISLANDS), ESPON & University of the Aegean, https://www.espon.eu/export/sites/default/Documents/Projects/TargetedAnalyses/EUROISLANDS/FinalReport_foreword_CU-16-11-2011.pdf

EURISLES (1999) The island regions and the price of intra-EU transport of goods, https://europeansmallislands.com/tag/eurisles/

EURISLES (2003) Islands transport and the EU, the situation in 2003. CPMR Islands Commission, https://europeansmallislands.com/tag/eurisles/

Eurostat (2011) Nomenclature of territorial units for statistics (NUTS), https://ec.europa.eu/eurostat/web/products-manuals-and-guidelines/-/KS-RA-11-011

Feick J (1992) Comparing comparative policy studies: a path towards integration? J Publ Policy 12 (3):257–286

Fischer F (2000) Citizens, experts and the environment: the politics of local knowledge. Duke University Press, Durham

Foweraker J, Landman T (2002) Constitutional design and democratic performance. Democratization 9(2):43–66

Gerring J (2004) What is a case study and what is good for? Am Polit Sci Rev 98(2):341–354

Gerring J (2007) Case study research: principles and practices. Cambridge University Press, Cambridge

Goode WJ, Hatt PK (1952) Methods in social research. McGraw-Hill, New York

Granovetter M (1985) Economic action and social structure: the problem of embeddedness. Am J Sociol 91(3):481–510

Grindle MS (2004) Good enough governance: poverty reduction and reform in developing countries. Governance 17(4):525–548

Gualini E (2018) Planning and the intelligence of institutions: interactive approaches to territorial policy-making between institutional design and institution-building. Routledge, New York

Hall PA (1993) Policy paradigms, social learning and the state: the case of economic policy-making in Britain. Comp Polit 25(3):275–296

Hjern B, Hull C (1982) Implementation research as empirical constitutionalism. Eur J Polit Res 10 (2):105–115

Hospers G (2003) Localization in Europe's periphery: tourism development in Sardinia. Eur Plan Stud 11(6):629–645

Howlett M, Ramesh M (1995) Studying public policy: policy cycles and policy subsystems. Oxford University Press, Toronto

Huber GP (1991) Organization learning: the contributing processes and the literatures. Organ Sci 2 (1):88–115

Hult KM, Walcott C (1990) Governing public organizations: politics, structures and institutional design. Brooks/Cole, Pacific Grove

Huntington SP (1968) Political order in changing societies. Yale University Press, New Haven/London

Ingraham PW, Joyce PG, Donahue AK (2003) Government performance: why management matters. Johns Hopkins University Press, Baltimore

ISTAT (2012) Censimento nazionale. Istituto Nazionale di Statistica, Roma, www.istat.it

ISTAT (2014) Italia in cifre. Istituto Nazionale di Statistica, Roma, www.istat.it

Jackman RW (1993) Power without force: the political capacity of nation-states. The University of Michigan Press, Ann Arbor

Kooiman J (ed) (1993) Modern governance. New government-society interactions. Sage, London

Koppenjan J, Klijn EH (2004) Managing uncertainty in networks: public private controversies. Routledge, London

Lawrence DH (1921) Sea and Sardinia. Thomas Seltzer, New York

Leeuw FL, Rist RC, Sonnischen R (eds) (1994) Can governments learn? Comparative perspectives on evaluation and organizational learning. Transaction Books, New Brunswick

Leonardi R (1995) Regional development in Italy: social capital and the Mezzogiorno. Oxf Rev Econ Policy 11(2):165–179

Lindblom CE (1990) Inquiry and change: the troubled attempt to understand and shape society. Yale University Press, New Haven

Lindblom CE, Cohen DK (1979) Usable knowledge. Social science and social problem solving. Yale University Press, New Haven

Lopez-Acevedo G, Krause P, Mackay K (eds) (2012) Building better policies. The nuts and bolts of monitoring and evaluation systems. The World Bank, Washington, DC

Lynn L, Heinrich C, Hill C (2001) Improving governance: a new logic for empirical research. Georgetown University Press, Washington, DC

Mathur N, Skelcher C (2004) The democratic performance of network governance: conceptual and methodological issues. International Conference on Democratic Network Governance, Copenhagen

Matland RE (1995) Synthesizing the implementation literature: the ambiguity-conflict model of policy implementation. J Public Adm Res Theory 5(2):145–174

Moyser G (2006) Élite selection. In: Jupp V (ed) The sage dictionary of social research methods. Sage, London, pp 87–88

Page E (1985) Political authority and bureaucratic power. Wheatsheaf, Brighton

Peters BG (1981) The problem of bureaucratic government. J Polit 43(1):56–82

Peters BG, Savoie DJ (eds) (2001) Governance in the twenty-first century: revitalizing the public service. McGill-Queens University Press, Montreal

Pierre J (ed) (2000) Debating governance. Oxford University Press, Oxford

Pierson P (2004) Politics in time: history, institutions and social analysis. Princeton University Press, Princeton

Plattner MF (2005) A skeptical perspective. In: Diamond L, Morlino L (eds) Assessing the quality of democracy. Johns Hopkins University Press, Baltimore, pp 77–81

Polsby N (1968) The institutionalization of the U.S. house of representatives. Am Polit Sci Rev 62 (1):144–168

Powell WW, Di Maggio PJ (eds) (1991) The new institutionalism in organizational analysis. University of Chicago Press, Chicago

Przeworski A (1995) Sustainable democracy. Cambridge University Press, Cambridge

Rhodes RAW (1996) The new governance: governing without government. Polit Stud 44 (4):652–667

Rhodes RAW (1997) Understanding governance: policy networks, governance, reflexivity and accountability. Open University Press, Buckingham

Rodríguez-Pose A (1998) Dynamics of regional growth in Europe: social and political factors. Clarendon Press, Oxford/New York

Rodríguez-Pose A (1999) Innovation prone and innovation adverse societies: economic performance in Europe. Growth Chang 30(1):75–105

Rose R (1974) The problem of party government. Macmillan, London

Rose R (1991) What is lesson-drawing? J Publ Policy 11(1):3–30

Rose R (1993) Lesson-drawing in public policy: a guide to learning across time and space. Chatham House, Chatham

Rosenau P (ed) (2000) Public-private policy partnership. The MIT Press, Cambridge

Sabatier PA, Jenkins-Smith HC (eds) (1993) Policy change and learning: an advocacy coalition approach. Westview Press, Boulder

Scartascini C, Tommasi M (2012) The making of policy: institutionalized or not? Am J Polit Sci 56 (4):787–801

Scartascini C, Stein E, Tommasi M (2013) Political institutions, intertemporal cooperation, and the quality of public policies. J Appl Econ 16(1):1–32

Scharpf FW (1989) Decision rules, decision styles and policy choices. J Theor Polit 1(2):149–176

Schumpeter JA (1943) Capitalism, socialism and democracy. Allen and Unwin, London

Scott RW (1995) Institutions and organizations. Sage, Thousand Oaks

Seawright J, Gerring J (2008) Case selection techniques in case study research: a menu of qualitative and quantitative options. Polit Res Q 61(2):294–308

Tebaldi M, Calaresu M (2009) Valutare la democrazia: introduzione all'analisi della qualità democratica. Aracne, Roma

Tebaldi M, Calaresu M (2015) 'Democra-city': bringing the city back into democratic theory for the 21st century? City Terr Archit 2(1):13

Tebaldi M, Calaresu M (2016) The pursuit of quality: assessing democracies...to save democracy? Open Journal of Political Science 6(1):1–9

Thelen K (2003) How institutions evolve: insights from comparative historical analysis. In: Mahoney J, Rueschemeyer D (eds) Comparative historical analysis in the social sciences. Cambridge University Press, New York, pp 206–240

Tommasi M, Scartascini C (2012) How (not) to produce effective policies: institutions and policymaking in Latin America. In: Santiso J, Dayton-Johnson J (eds) The Oxford handbook of Latin America political economy. Oxford University Press, Oxford, pp 263–284

Yanow D (1996) How does a policy mean? Interpreting policy and organizational actions. Georgetown University Press, Washington, DC

Chapter 3
Institutionalisation of the Governance System and the Quality of Urban Regeneration Policies in Sardinia

Abstract This chapter is devoted to the empirical analysis of the urban regeneration policy for the typical case study of the cities of Sassari and Alghero in the Mediterranean island of Sardinia, Italy (1995–2006). In the first section of the chapter, we present the urban regeneration policy in Italy and Sardinia, addressing the relevant actors, objectives, resources and instruments embedded in the policy. In the second section, we test the degree of institutionalisation and the quality of the policy in the policy area within the selected case study. For this purpose, we highlight several analytical dimensions of the institutionalisation of the governance system as well as the quality of the territorial policies. In the third section, we perform an overall assessment of the case study, paying particular attention to evaluating the institutionalisation of the governance system and the quality of each observed territorial policy.

Keywords Urban regeneration policies · Sassari · Alghero · Sardinia

3.1 Urban Policies and Urban Regeneration Policies

Urban policy could be interpreted as a 'policy directed to cities', to draw a distinction between non-urban or rural, regional and national policies. The term indicates only the spatial range of the policy and says nothing about the content nor about the authority responsible for such policies. In general terms, urban policy can be then described as the whole set of government measures at different administrative levels—European, national, regional or local, which are directed to cities (van den Berg et al. 2018, 12). Aiming to take a step forward defining urban policies, we highlight below three of the main characteristics of urban policies which are nowadays considered by scholars to be most common in contemporary democracies:

1. The first characteristic concerns the objectives aimed at by these heterogeneous policies. These objectives are in fact related to two different kinds of solutions to collective problems (Stoker and Mossberger 1994; Gualini 2017): the design and

© Springer Nature Switzerland AG 2021

M. Tebaldi, M. Calaresu, *The Quality of Territorial Policies in Europe's Periphery*, Urban and Landscape Perspectives 22,
https://doi.org/10.1007/978-3-030-62647-1_3

planning of urban physical spaces, also known as *urban planning policies*, and the governance of the economic, political, social and cultural development of the city, also known as *urban policy agendas*.

2. The second characteristic concerns the decision-making configuration of urban policies. The governance of the city takes place through an extremely complex and composite network of interconnections (Wolman and Goldsmith 1992; Andrew and Goldsmith 1998; Christensen 1999), multiple initiatives and programmes, coalitions of players and mobilisations (DiGaetano and Strom 2003). This is although that the territorial scale and the target social groups of these policies and their effects are limited and restricted (Pierre 2005). More precisely, one of the peculiarities of these policies nowadays is that they tend to have a multilevel governance configuration since the sets of plans and courses of action undertaken and implemented involve interaction between different levels of governance and government (John 2001) and interaction between public and private actors, in more or less formalised partnerships (Kaufmann and Sidney 2020).
3. The third characteristic, closely linked to the first two, emphasises the cross-sectoral nature of urban policies and how they often engage with multiple policy subsystems within individual decision-making processes, such as collective public services, social services, urban and town planning, public works and productive activities (Leat and Setzler 2002; James 2015). Urban policies today, in fact, involve know-how which only a few decades ago was regarded as implicit or marginal to 'classical urban planning' but which is now characterised by 'the transcending of disciplinary barriers, conceptual migrations, common languages, and concepts linked to new modes of thinking about dwelling space' (Maciocco and Clemente 1996).

In the last two decades (2000–2020), the theme of urban regeneration has represented the principal innovative element within urban policies (Maciocco 2007a, b). The term urban regeneration—perhaps because of its multidisciplinary and interdisciplinary use—is a somewhat ambiguous term which occurred first in the United Kingdom (later spreading to other parts of the world) and which is used in multiple senses to describe a diverse range of scientific approaches or public policies (Saccomani 2016). However, all are generally focused on economic development and physical and architectural improvement through assistance programmes for disadvantaged social groups (Saccomani 2004).

Many authors have highlighted the substantial differences in the use of the term—especially in a diachronic sense—by speaking of 'urban regeneration periods' or 'generations of urban regeneration policies' (Bighi 2016; Saccomani 2016).

The first period has been located at the end of the 1970s, beginning in the United Kingdom, and has been associated with the so-called flagship projects; large-scale projects mostly under private management oriented towards the *urban renewal* of specific sites 'according to a mainly real estate model [. . .] unconnected to a holistic vision of the city' (Bighi 2016, 115). In line with the neoliberal economic model of urbanisation, the underlying assumption of these projects was that prosperity would *trickle down* to the most economically and socially disadvantaged areas which had been excluded from the direct process of urban reconfiguration. More generally, beyond the borders of the United Kingdom, the use of flagship projects was a consequence of the deindustrialisation and dismantling of the great urban industrial areas which took place in many European cities, as a result of the socio-economic transformation which began precisely at the end of the 1970s. Subsequent research has shown how these projects have paradoxically brought about greater socio-economic polarisation and lower community participation (Saccomani 2016).

The second period can be located about a decade later, beginning in the 1990s, and has been associated with a more integrated vision oriented towards *urban regeneration*, with projects designed to deal simultaneously with economic, physical, social and environmental problems (Saccomani 2016), in historical town centres or Old Towns, as well the peripheries or semi-peripheries (Guiati 2008) of many post-Fordist cities (Bagnasco 1990). Some of the problems addressed were the degeneration of the housing stock and the spaces external to dwellings, monofunctional character of dormitory suburbs, social hardship, low transport accessibility and isolation (Sbetti et al. 2013). In the same years, another theme emerged, challenging 'the widespread assumption that the [compact] city is the only thing of interest' (Maciocco 2007a, 30), with a new emphasis on the environmental implications of urban planning (Maciocco and Camarda 1991) and on environmental landscape planning (Maciocco 1991)—with attempts to limit the consumption of non-urbanised surface areas ('empty landscapes'), as opposed to other more densely developed areas (Maciocco 2007a, b, 2008), and the 'fringe areas' (Maciocco and Pittaluga 2001). The most common project model in this second period was very close to the so-called *programmi complessi* (complex programmes) which were also developed in Italy. The predominance of the private sector was considerably reduced compared to the first period, with integrated place-based and area-based type policies (Andersen 2001), with a greater emphasis on the role of the community, on partnerships between public and private sectors and voluntary associations and in particular on active participation, expressed through the direct or indirect involvement of the local population (Saccomani 2004; Allulli 2011).

The third period which has been located between the end of the 1990s and the second decade of the 2000s has been associated by scholars with the role of the EU (Couch et al. 2008) and its various institutional articulations: the European Commission, the European Parliament and the Council of Europe. Even though the EU treaties do not explicitly take into account intervention in urban policy by the EU institutions, in these years, the EU has faced an increasing pressure to develop a European urban planning policy, since that more than half of the population in OECD countries (53%)

live in urban areas (OECD Territorial Reviews 2006) and European cities were considered 'places of social and cultural integration, [. . .] sources of economic prosperity and sustainable development, and [. . .] the bases of democracy' (European Commission 1997, 3). The so-called European urban regeneration policy came into being with the publication by the European Commission of two documents: *Sustainable Urban Development in the EU: A Framework for Action* in 1998 and *European Spatial Development Perspective* in 1999 (Novembre 2011). The Commission in this way established a series of objectives to promote and increase economic prosperity and employment in towns and cities, equal opportunity, social inclusion and recovery of neighbourhoods, to improve the urban environment, and contribute to good urban governance and increased participation by citizens and local stakeholders (Del Genio 2007). Although many authors believe that some of these objectives have become purely rhetorical over time (Governa et al. 2009; Saccomani 2016), they have still contributed to making the methodologies and approaches followed by various EU countries more homogeneously and scientifically focused. More concretely, EU interest was reflected in the introduction of large-scale programmes and new funding schemes, especially within the 1994–1999 Structural Funds framework. These programmes were innovative not only in terms of the forms of financial support but especially in terms of the request for detailed plans of action featuring financial, social and city planning sustainability. Furthermore, a new project evaluation process was introduced in those years at the EU level, based on the documentation and evaluation of an 'emergency' (defined as a necessary condition for a successful funding request) in order to assess the project in at least some of its fundamental aspects such as feasibility and project sustainability, capacity to stimulate the development of new initiatives aimed at the economic development of urban areas, ability to involve economic and financial private stakeholders in the process and capacity to affect different levels of urban organisation and various problems concerning the local community (Savino 2006).[1]

In search of a possible further, more specific definition of urban regeneration policies—without any claims to exhaustiveness in considering the complex semantic and the several disciplinary fields related with this concept—we can state that these

[1]Apart from the analytical scheme proposed above, other authors (Moulaert and Mehmood 2020) also refer to the cyclical nature of urban regeneration policies, stating that in the last 20 years, EU countries have exhibited a *pendulum movement* in urban regeneration policies, moving from a neighbourhood-focused, socially community-oriented policy approach in the 1990s to a more city-wide, economy and real estate-driven policy in the 2000s, before in the last few years restoring to the centre of urban regeneration policies the notions of social innovation and attention to the neighbourhood, although now situated within a multi-scalar context (Gualini 2006). By way of example, while the European programmes of the 1990s like *Poverty III*, *Urban Pilot Projects*, *Urban I* and *Neighbourhoods in Crisis* assigned a central role to urban neighbourhoods—to the point of considering them the most appropriate territorial level at which to initiate strategies for combatting social exclusion and poverty (Atkinson and Carmichael 2007)—with the advent of the new millennium, this territorial level became subject to debate, with the focus of urban regeneration policies now redirected back to the city (Tebaldi and Calaresu 2015) and regional levels (Gualini 2004).

policies nowadays are marked by the renovation and redevelopment of the urban environment in various forms: the *soft* form (beautification) which is primarily expressed through the creation of places and spaces, buildings, streets, squares or other kinds of street furniture, favouring the construction or the rediscovery of local identity, the improvement of commercial and tourist activities and the economic, cultural and green areas and collective play areas, and the *hard* form, that is, the destruction and rebuilding of entire sections of the city. The common feature of these two types of development is *urban regeneration* (Karrer et al. 1998), defined by Evans and Shaw (2004, 4) as:

> The transformation of a place (residential, commercial or open space) that has displayed the symptoms of environmental (physical), social and/or economic decline. What has been described as: breathing new life and vitality into an ailing community, industry and area [bringing] sustainable, long term improvements to local quality of life, including economic, social and environmental needs.

So urban regeneration policies could refer to any actions directed towards areas often located in older urban peripheries or semi-peripheries or Old Towns often affected by major issues related to poverty, degradation, weak service networks and the presence of processes of segregation and social stigma (Chiodini and Milano 2010). The quality of life in these areas is often a result of the interaction of many environmental, economic and social–cultural aspects. Consequently, working through several converging purposes, such as architectural and urban redevelopment, environmental improvement, the boost to the local economy, the fight against the marginalisation of groups exposed to social risk factors and promoting participation, these policies aim to achieve a reasonable level of urban and life quality (Mela 2006).

The idea of the *renovatio urbis* project is based on a specific set of plans and processes, limited and dedicated to strategic locations. These plans aim to change—possibly in radical ways—the function, role and image of entire parts of the city or the entire city itself (Secchi 2005). In many cases, they are actions directed at the whole urban organisation; or instead, these plans and processes may be projects or measures focused on more limited urban spaces or individual buildings, in an attempt to create a strong and significant urban *sign* within the city fabric. In general, urban regeneration policies have sought to transform existing urban areas to redevelop them while trying to maintain their original purpose (Savino 2006).

3.2 Urban Regeneration Policies in Italy: Players, Objectives, Resources and Tools

According to scholars in this area, it is difficult to establish with any certainty when urban regeneration first became part of the Italian political agenda. Up until the beginning of the 1990s of the last century, the urban question in Italy obscured any issues related to urban planning, given that no explicit necessity had been identified within the political agenda for a policy for the city oriented towards facing the challenges posed by the urban condition in all its complexity (Allulli 2011).

The forces in play at the national and local level up to this point were directed towards breaking free from the National Urban Law 1150/1942, as well as full implementation of Law 75/1967. Other European and North American contexts had strategies and programmes at various territorial scales as well as specific institutional instruments (Pasqui 2011) from at least the beginning of the 1970s. The reason for the delay in Italy, according to some authors, was that for a long time there was no institutional definition of local government which took into account geographical areas destined for large-scale urban and territorial plans. Other authors highlight how the delay in the arrival of the urban question on the national political agenda can be ascribed to two other types of factors, both of them institutional in nature (Allulli 2011, 8): 'On the one hand, the rules of the game which had been decided during the foundational phase of the Italian Republic had identified the regions as the means for decentralising governing functions, principally for reasons of political opportunism; on the other hand [. . .] the high degree of consolidation of city governments seems to have brought about a "taken for granted" approach to the role of the city, and apart from a few exceptions, brought about a substantial absence of specific attention to the city within the debate over institutional innovation'.

The absence of urban regeneration policies in Italy (Cremaschi 2002) began to be addressed, thanks to a series of national legislative measures designed to renovate the built heritage. For example, with the National Law 457/1978 (the so-called Ten-Year Housing Plan), considerable resources were dedicated to renovation programmes for the first time. Despite the size of the resources assigned, the 1978 Law did not, however, lead to a widespread process of urban regeneration. The inadequacy of the regulations on renovation came about due to the sector-specific nature of the legislative provisions, so that interventions were limited to the housing sector, divided into many small-scale initiatives, and never managed to take on a proper urban dimension. For this reason, some authors prefer to consider the establishment of the Ministry for urban areas in 1987 as 'symbolic of the appearance of the urban question on the national political agenda and the government agenda' (Novembre 2011, 127). On paper, in fact, the newly created Ministry was given the roles of promotion, networking and coordination of urban policy initiatives at the national level, with the aim of intervening in situations in which local communities were not immediately responsible. However, as a result of the national political crisis of the early 1990s known as *Mani Pulite* (Clean Hands), this Ministry was abolished (1993), without having been able in the meantime to make any headway in addressing the fragmentation of national government and governance assets by imposing order on the multiplicity of actors and networks which often overlapped in terms of their responsibilities. This led to criticism of the urban policies under consideration from the point of view of efficiency and effectiveness.

Following the abolition of the Ministry mentioned above, there emerged an even more evident need at the national level to promote publicly and privately financed integrated programmes across a range of contexts (from housing to services and businesses) through a plurality of public–private operators. These integrated-type programmes had already been partly taken into consideration in the resolution of the Interministerial Committee for Economic Planning in relation to the 1988–1989 biennial programme for the Ten-Year Housing Plan and, later, within the ambit of Italian Law 203/1991. This law provided under Art. 18 for an extraordinary residential housing intervention to support the mobility of state personnel. But it was during the 10-year period 1990–2000 that urban regeneration policies in Italy were able to make a qualitative leap through the introduction at various levels of government of innovative solutions in terms of planning, programming and technical equipment. In particular, a series of legislative measures was introduced to support urban regeneration through the promotion of the complex programmes already mentioned, which under Law 179/1992 were divided up into integrated privately initiated programmes (Art. 16) and Urban Regeneration Programmes (Contardi et al. 1999) promoted by the Ministry of Public Works. Later, Law 493/1993 initiated programmes under Art. 11 for urban recuperation, under regional responsibility and 'at the service of the public built heritage'.

All of the programmes mentioned so far refer to the institutional programme agreements and the service conferences (see Chap. 1, notes 12–13, for further specifications) implying the possibility of coordinating public funding and a faster decision-making process. In the background, there was also the possibility of requesting substantial modifications to urban planning regulations in terms of zoning and building index, which in many cases turned out to be a greater stimulus for urban regeneration than public funding itself. Financial Law 662/1996 provided resources for innovations such as the Neighbourhood Contracts previously mentioned, instruments for urban regeneration which took into account European and international recommendations and represented a type of integrated programme which not just reflected building and urban planning elements but was also open to considering social and employment questions.

The establishment of a new paradigm for urban regeneration policies was favoured by certain political–institutional factors, among which we can mention the Single Text on Local Autonomies (Law 142/1990), the laws reforming public administration (the so-called 'Bassanini laws'), the establishment of instruments for negotiation procedures (such as the Territorial Pacts and Area Contracts) introduced under the Financial Law of 1996 but above all the introduction under Law 81/1993 of the direct election of city mayors. From that moment on, mayors took on the administration of complex processes, including from the point of view of institutional responsiveness and accountability (Tebaldi and Calaresu 2009), as well as transition of business production; physical transformation of the city, with particular attention to the question of the repurposing of abandoned or underutilised spaces; support for human capital through the strengthening of research and higher level and professional training; the provision of infrastructure for local production systems and the whole urban region; and contribution to cultural life and attracting special events (Lees 2008; Pasqui 2011).

Within this paradigm shift, the regeneration of Old Towns has played an important role. The reuse, upgrading and redevelopment of the existing city have been considered a new frontier of urban development[2] (Padovani 2000; Savino 2006). This change in the aim of urban regeneration policy has been characterised by the slowing down of urban expansion and increasing attention to the areas already built within the cities (Savino 2006). In this perspective, it is also possible to explain why and how urban regeneration policies can be closely linked to other policy areas, in particular, those related to the organisation of sports and cultural events which can have a major social and economic impact at the territorial level[3] (Guala 2007). Hosting a major event for a city requires, in fact, the construction of new urban infrastructure, logistical support areas and media hubs. Therefore, the whole public transport system at the local, regional and national levels has to be reorganised, several unused urban areas and industrial areas are recovered, and the hospitality system is modernised (Guala 2007). Furthermore, such interventions often result in buildings, monuments or public works, which become or may become the everlasting symbols of the event itself.[4]

The functional characteristics of these projects are fairly homogeneous and can be divided into three major categories. Transport and communication networks are

---

[2]In the past, the increase in urban population and the changing needs of the community from the social, economic and cultural standpoint dictated by the advent of the Fordist production model, as well as the change in the use of living and working spaces, shifted attention towards the peripheral areas of the city, causing a virtual drainage of city centres. Over time, the patent and marked depopulation of old towns, which in some cases resembled a desertification process, has been characterized by some concurrent factors: (a) the physical and social degradation of residential properties, which led the inhabitants to abandon them and to move to the suburbs, (b) the progressive increase in car traffic volume and the issues related to the accessibility of the city centre and (c) the fact that commercial activities and services moved from the centre to the outskirts of the city mainly due to their higher degree of accessibility. The old town was abandoned because it no longer met the new financial and social requirements. This phenomenon is called by geographers *de urbanization*. The city centre began to lose part of the population due to the factors listed above as well as changes in the industrial process which have reduced the demand for labour by companies. Peripheral areas, which used to be marginal within economic and demographic development, became more attractive, leading to a diffuse urban expansion (Mela 2006).

[3]The events mentioned in academic literature include international fairs and world exhibitions, but especially major sports events, such as the World Cup or the Olympic Games (e.g. for the latter, see Nel·lo 2010). It is quite clear that the implications of these events go beyond sports itself (they are also extremely important in terms of engaging other factors of economic, social and cultural significance). There are other events which are also of considerable importance in the political sphere, such as organizing the G7 or G8 meetings or, within the cultural sphere, events such as the European City of Culture (Guala 2007).

[4]This is the case for the Eiffel Tower, built for the 1889 Universal Exhibition, on the 100th anniversary of the French Revolution. It was supposed to be a temporary structure, but it has become one of the symbols of Paris, thus going far beyond the function that it was originally supposed to play (Guala 2007).

most important; the need to improve the mobility of people and facilitate the exchange of information is nowadays essential, and therefore, metropolitan railways, parking lots and urban highways are clearly pivotal. A second category consists of construction projects aimed at developing the tertiary public sector and/or connected to economic production. The former group includes new headquarters for the national and local public authorities, especially for traffic-generating functions, such as courthouses or universities. The latter group includes technology parks and trade facilities, such as fairs and exhibition venues. The third group of functions provided for new projects concerns leisure and entertainment activities, including cultural and sports facilities, theatres or metropolitan parks (Dente 1990). Urban regeneration processes aim at strengthening competitive territorial factors; the dynamism of the city is secured by environmental quality, infrastructures, educational system, cultural heritage and urban quality (Galdini 2008).

Because of the complexity of the new policies, the role of the city mayors has been supported by a multiplicity of national public actors (Allulli and Tortorella 2013). First of all, the Italian Ministry of Public Works, together with the Ministry of Infrastructure and Transport, began to plan projects and define policy instruments in line with the provisions of the EU. This activity was expressed both in allocating financial resources and in the introduction of new project protocols, leaner procedures and preliminary evaluation processes to justify planned actions and related public investments. The Ministry of Public Works and the Ministry of Infrastructure and Transport were ‘complemented’ by the Ministry of Finance, which funded Territorial Pacts using a bottom–up approach (Savino 2006, 103). The ministries thus became promoters of new actions, which complemented EU funding schemes and became pivotal for the urban transformations of Italian cities throughout the 1990s. That decade was a period of intensive experimentation with new modes of action in the regeneration of degraded urban areas. The awareness of the risks of social and physical degradation in many neighbourhoods, the need to remedy imbalances caused by rapid growth during decades of uncontrolled expansion and the need to create new opportunities for urban economic and social development prompted the definition and implementation of public policies aimed at redeveloping and regenerating urban territories.

Secondly, we can mention the General Directorate for Territorial Development, Planning and International Projects (DICOTER hereafter) of the Italian Ministry of Transport and Infrastructure. DICOTER has been responsible for the promotion and implementation of urban regeneration policies through complex programmes (e.g. Urban Regeneration and Sustainable Territorial Development Programmes; see Chap. 1, notes 12–13, for further specifications), through its role in collaboration and technical assistance in the implementation of European projects, such as Urban and Urban II programmes, and Urban Pilot Projects (Palermo 2002a, b; Palermo and Savoldi 2002), created as part of specific cohesion policies (Del Genio 2007). The introduction by DICOTER of new planning practices and new urban instruments has provided an important opportunity in Italy for the regeneration of urban areas affected by social decline and economic marginality, as well as for the regeneration of decommissioned urban industrial areas, and finally, but not in terms of importance, of the modernisation of the city in terms of the development of new strategic functions (above all, tourism and technological innovation).

Due to the new opportunities created by the new EU funding schemes (Cremaschi 2002), other governmental bodies have shown greater dynamism and proactiveness, such as the Department for Development and Cohesion Policy of the Ministry of Economics (today located within the Ministry for Economic Development), which has been one of the leading agents in planning the disbursement of European Structural Funds.

In the Italian case, Structural Funds required that urban policies were embedded in regional policies, with the regional governments setting priorities and resources and selecting cities within their territories which could benefit from the funding. According to this scheme, the European Commission urged regional governments in Italy to delegate to the municipalities the direct management of funds and Member States to set their strategy for urban development in the context of harmonisation and competitiveness. In this context, the market's renewed interest in urban values is stimulated by chance for redeveloping the city and encourages renewed interest by public institutions. The latter had extensive real estate properties resulting from, for example, industrial areas, which became investment opportunities for sale and/or urban regeneration. The urban regeneration era in Italy was characterised by the proliferation of public–private project financing schemes, new planning instruments and negotiation procedures that affected almost all the towns and cities in the country. Furthermore, the southern part of Italy (included in the EU Objective 1 regions) used financial territorial integrated projects provided within the framework of the 2000–2006 Structural Funds (Savino 2006; Del Genio 2007).

To the list of national and local actors outlined above should be added the regions, which have been protagonists in Italy since the beginning of the 2000s in urban regeneration policies, as a consequence of the regionalisation of government practices and planning instruments on a local territory as well as national and above all European level. The process of change initiated by the regions is significant because from the regulatory point of view it represents a new way of using the resources of the territory and planning urban and territorial regeneration interventions and actions (Novembre 2011).

3.3 Urban Regeneration Policies in Sardinia: Regional Planning Tools

As already mentioned above, urban planning legislative activity by the Italian regions did not enjoy the support of a national framework law. The competences in this area were transferred from the central state to the regions in two ways: the Italian 'special statute' regions, including Sardinia, acquired them through their constitutional laws at the time of their establishment,[5] whereas Italian 'ordinary' regions acquired their competences in the field of spatial planning in the framework of Presidential Decree 616/1977.

Urban development planning is synonymous with territorial governance intended not only as processes of definition and structuring of the landscape but also processes such as the launch, management and implementation of transformational plans and programmes according to definite times and modalities, inclusive of evaluations of the impacts. Regional urban planning (a) defines the regional policy objectives, (b) defines the constraints for planning undertaken by the lower governmental territorial levels, (c) is mostly directed to other political and administrative institutions and has no binding nature for citizens, (d) is the reference point for any special and sector-specific planning, and (e) helps define the objectives of the regional economic development plan (Indovina 2012).

The history of landscape planning in Sardinia has been particularly beset by difficulties. The impulse for the development of the Regional Landscape Plan resulted from the legislative constraints of Article 1 (bis) of the so-called Galasso law (Law 431/1985) that provided for the obligation to 'draft landscape plans or urban-territorial plans paying attention to specific territorial values' through which the territory could be protected and promoted. The main regional legislative frameworks governing landscape, city and territorial planning are Regional Law 45/1989 and Regional Law 8/2004, which amended the first state reform in landscape protection (Legislative Decree 42/2004).

Regional planning instruments, according to Art. 3 of Regional Law 45/1989, govern the use and protection of the territory. They consist of landscape territorial plans,[6] directives and constraints and territorial planning schemes. The legislative framework was amended by Regional Law 8/2004 that specified that the *Piano Paesistico Regionale* (Regional Landscape Plan, hereinafter referred to as RLP) is the regional territorial framework of reference, that is, the conceptual and legislative reference plan of the whole regional and subregional planning system (Sanna 2011, 25). Therefore, the RLP was a tool to protect the entire regional territory, which not only provided information and descriptive content but also embodied propositional and regulatory elements. The RLP marked the shift from a system tied down by territorial landscape management as the main tool for landscape conservation to a

[5]The Autonomous Region of Sardinia acquired those competencies via Constitutional Law 3/1948.

[6] Under Art. 10 of Regional Law 45/1989, these plans must include the morphological–historical analysis of the territory and the structure of the landscape, the identification of landscape scenarios and various levels of use and the definition of territorial environments, including the abovementioned scenarios and compatible criteria of use. In particular, they must include the areas where the full preservation of natural, historical–morphological characters and their coming together must be guaranteed; the settings for which are allowed transformative interventions by specifying the limits, criteria, as well as the maximum volumes of construction; scopes for which restoration and environmental recovery are necessary, specifying their characteristics; the definition of territorial areas for which the transformations of the territory and the active intervention of conservation and restoration depend on the adoption of provincial plans; and the criteria and rules for implementation.

system that considered regional planning as the tool to enact the constraints imposed by the national legislation. It also establishes additional protective measures to be implemented through coordinated action with local stakeholders, which carry out major administrative functions related to urbanisation and construction. The RLP may also provide for the regeneration of compromised landscapes. In that case, however, the active intervention of other stakeholders, such as the municipality, the province and other interested bodies and individuals, is also required for the implementation of restoration and reclamation plans or of programmes for the protection or conservation of safeguarded building and areas, the implementation or management of which could be attributed directly to local authorities (Sanna 2011).

In addition to defining its RLP, the Sardinia Region, through the enactment of the Regional Law 29/1998, tried to emphasise the issue of Old Towns, defined by the law as 'urban conglomerations that retain the signs of ancient constructions and its original housing, economic, social, political and cultural functions through the territorial organization, urban planning, and buildings'. The main purpose of the regional law is the development of the real estate resources available, to be achieved together, however, with the limitation to the erosion of territorial resources. According to the above-mentioned regional law, this twofold objective had to be attained by means of the 'recovery, redevelopment and reuse of old city centres and minor historical settlements by protecting their social-cultural, historical, architectural, urban, economic and environmental values'. The objectives of the lawmakers were implemented through the creation of a census of Sardinian Old Towns containing all the buildings to be potentially included in the protection and regeneration effort. This census was the general framework of reference for the regional planning actions.[7] With the aid of the census tool, the multi-annual programme of the Old Towns was drafted as the main implementation tool of urban regeneration planning, including the municipal funding schemes. The multi-annual programme may contain integrated programmes[8] or specific urban redevelopment actions.

[7] According to Art. 5 of Regional Law 29/1998, it contains the municipalities that are characterized by the presence of buildings and constructed areas which present urban and architectural, cultural and environmental value, typological connotation or aggregation, and therefore, they are eligible as regional fund beneficiaries for urban and construction rehabilitation. The requirements for the municipalities to be included in the list are as follows: (a) a consolidated and enclosed urban–real estate property as evidenced by the historical land registers before the year 1940 currently showing (1) a connective urban fabric, consisting of roads, squares and public venues, essentially unchanged, and (2) buildings mostly retaining the characteristics of historical settlements, (b) mostly homogeneous construction and technological characteristics, (c) homogeneous and widespread architectural elements and (d) an abstract of the current town planning instrument containing the perimeters of the urban area concerned.

[8] Under Art. 9 of Regional Law 29/1998, the integrated programs of Old Towns are characterized by (a) a dimension sufficient to affect the urban reorganization of city centres; (b) the presence of a multiplicity of functions; (c) the integration of different types of intervention, including works on infrastructures; (d) cooperation of public and private stakeholders; and (e) a set of public and private financial resources.

The Sardinia Region also took action by preparing the Regional Operational Plan (ROP hereafter) by providing for a special axis called 'cities'. The latter, being inspired by the principles contained in the Community Support Framework (CSF hereafter), defined as the overall objective:

> To improve the functional articulation and the quality of the urban system of Sardinia through the definition of the role of cities in their regional context, and in particular to: redevelop the social-economic, physical and environmental contexts of neighbourhoods and urban areas, improving their living conditions and creating suitable conditions for the development of entrepreneurship; promoting the launch of new services for the benefit people and businesses, boosting the competitiveness of territorial economic systems; combat social marginalization and promote social confidence recovery processes.

The regional strategy is based on three different levels of action implemented taking into account the specific characteristics of the territory; they are divided into:

(a) Strengthening services of excellence in the main urban settings and their peripheries, through a coordinated set of locational interventions of non-ordinary functions (innovative activities, advanced training centres, global network hubs) and the recovery of degraded neighbourhoods mainly through the expansion of social services for individuals and communities.
(b) Adjustment of services in medium-sized cities, which support the dynamism of local productive systems, enriching and strengthening urban functions in support of the territory. Those medium-sized cities carry out political and administrative functions, and their populations are large enough to require the provision of significant social–economic services in the territory of reference.
(c) The revitalisation of small-sized cities and towns which, despite lacking the above-mentioned characteristics, plays a major role in production activities and local systems for the development of internal areas.

The implementation of the regional strategy took place according to the size of the urban area of reference; therefore, the axis provided the actions to be carried out in the so-called 'main' urban areas and medium- and small-sized urban areas, in addition to actions which were common to all urban settlements, regardless of their demographic size. The plans and programmes provided for the main urban areas included but were not limited to the upgrading, redevelopment and reorganisation of urban green spaces, cultural services and social services to individuals and the community, implemented through urban redevelopment programmes (e.g. Neighbourhood Contracts; see Chap. 1, notes 12–13, for further specifications) in degraded districts (e.g. Old Towns and urban peripheries) and in abandoned areas. The plans undertaken in medium- and small-sized cities and towns involved actions, such as strengthening and adjusting the structure and management of services to support the dynamism of local productive systems and enriching and strengthening at the same time the spectrum of urban functions supporting the territory.

Among the actions common to all urban settlements in the Region, the following plans and programmes are included: providing suitable training to bureaucrats to manage the new models of urban governance, facilitating the establishment of a regional observatory dedicated to urban transformation for a regional-scale analysis of cities and city networks and adjusting the planning tools of urban transformations, including through the preparation of pilot projects.

The Region of Sardinia also regulated the drafting of the Municipal Urban Plan (MUP hereafter) to be implemented by the municipalities through Regional Law 45/1989. The MUP sought to ensure a balanced expansion of urban settlements in line with regional regulations and guidelines. The MUP regulated the use of agricultural land and parts of the territory allocated to the development of tourism and industrial–artisanal production; it laid down the rules for the recovery and use of existing buildings; it regulated the provision of social services and infrastructures within the municipalities.

The protection of Old Towns was defined by the introduction of the so-called Urban Recovery Plans (RP hereafter; see Chap. 1, notes 12–13, for further specifications) within the above-mentioned Regional Law 45/1989. The RP was usually drafted by the municipality, which also used it to invest the funds allocated by the regional government, starting the recovery of both degraded publicly owned and privately owned housing estate through expropriation or an agreement with the owners (if the property was of paramount public interest) (Karrer et al. 1998, 74). The RP is a comprehensive urban restoration plan aimed at improving the conditions of degraded areas, with interventions on both buildings and urban organisation (Indovina 2012, 196–198).

3.4 Analysis of Two Typical Case Studies

This section of the work presents the empirical analysis results on the institutionalisation of the governance system of urban regeneration policy and its quality in the two typical case studies of Sassari and Alghero, the two largest cities in the northwest of Sardinia (Fig. 3.1). As already mentioned in the second chapter of this book (Chap. 2), the analysis covers a period, ranging between 1995 and 2006 in both cases.

As illustrated in the following table (Table 3.1), the empirical analysis of each case is conducted by examining the following: firstly, the independent variable, namely, the institutionalisation of the governance system, aiming to test the degree of its three dimensions empirically: (a) *degree of functional specialisation and interdependence maps among the actors of the dominant coalition*, (b) *degree of congruence of cognitive maps within the dominant coalition* and (c) *degree of structuring and impermeability of gatekeeping*, and, secondly, the analysis of the

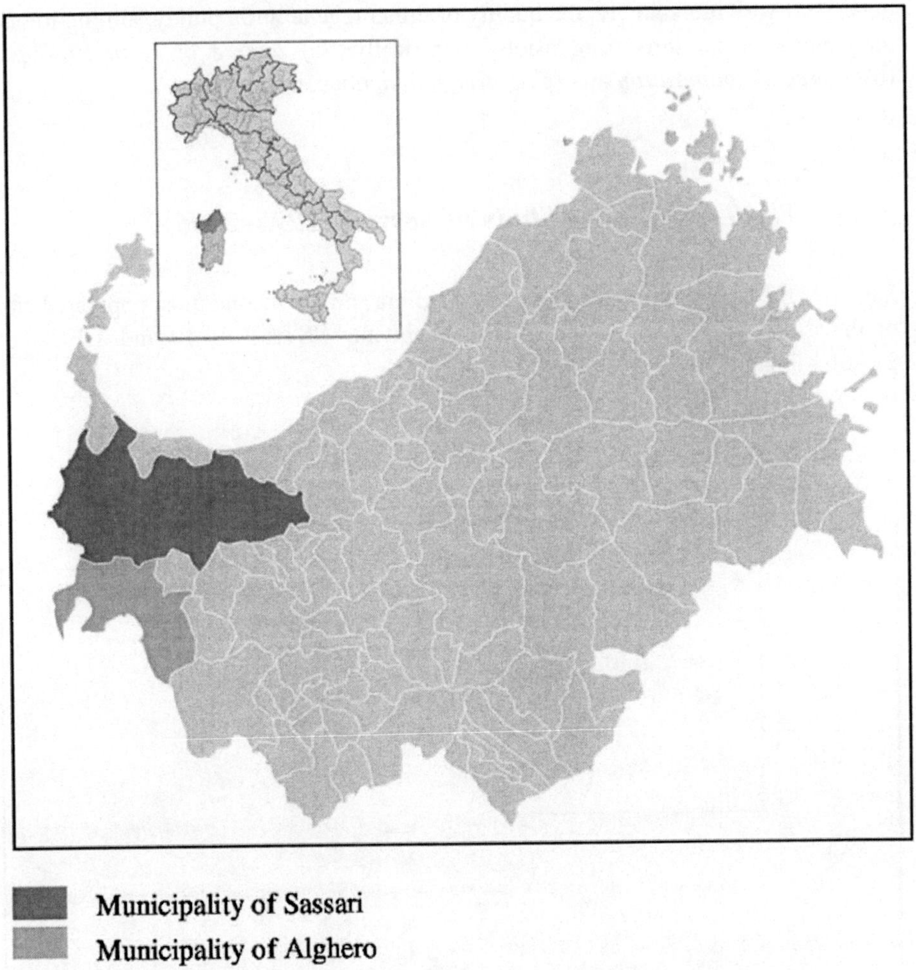

Fig. 3.1 The municipalities of Sassari and Alghero

Table 3.1 Variables of the research and their analytical dimensions in the case studies of Sassari and Alghero

Independent variable	**Dependent variable**
Institutionalisation of the governance system	*Quality of the urban regeneration policy*
Analytical dimensions	**Analytical dimensions**
(a) Degree of functional specialisation and interdependence maps among the actors of the dominant coalition	(a) Degree of responsiveness
(b) Degree of congruence of cognitive maps within the dominant coalition	(b) Degree of cooperation
(c) Degree of structuring and impermeability of gatekeeping	(c) Degree of functional congruence

dependent variable, namely, the quality of urban regeneration policy, aiming to test the degree of its three dimensions empirically: (a) *degree of responsiveness*, (b) *degree of cooperation* and (c) *degree of functional congruence*.

3.4.1 The Typical Case Study of Sassari (1995–2006)

Sassari is the largest city of northwest of Sardinia and the second most populated city of the island (Fig. 3.2), with 126,870 inhabitants (ISTAT 2019) and a FUA of 215,364 inhabitants (Eurostat 2020).

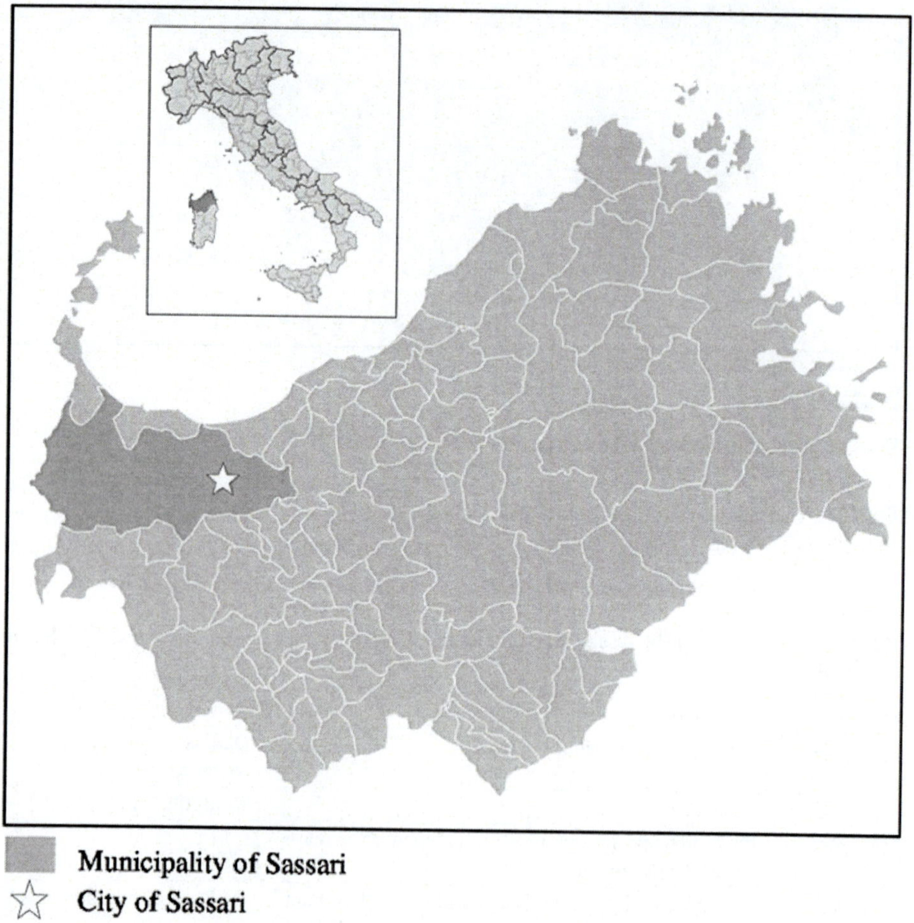

Fig. 3.2 The municipality and the city of Sassari

3.4.1.1 The Institutionalisation of the Governance System

Degree of Functional Specialisation and Interdependence Maps Among the Players of the Dominant Coalition

As far as the institutionalisation of the governance system is concerned, the first dimension to be analysed relates to the (a) degree of functional specialisation and interdependence maps among the actors of the dominant coalition.

The regeneration of the Old Town of Sassari began in the second half of the 1990s and, more specifically, during the 5-year term of Mayor Anna Sanna, who was in office between 1995 and 2000. The policy was first funded by the Urban Regeneration Programme (see Chap. 1; notes 12–13, for further specifications) and later by the CSF. The plan to redevelop the area using those funds was a choice of the then ruling centre-left municipal government:

> Works in the Old Town of Sassari started in 1996 under a national call for urban regeneration plans. The then Mayor Anna Sanna submitted a project coordinated by Professor Maciocco and his team of architects. We focused on a regeneration plan for the Old Town. The project was approved. We were planning to recover some historical buildings, such as the San Pietro infirmary. We regenerated a building that had been burned down and abandoned, and it hosted the headquarters of the Department of culture of the Municipality of Sassari. Some minor works were carried out, including at the [. . .] *Palazzo Civico*. At the same time, I was working on a project involving the Ministry of the environment to complete the CSF. The program had ended in 1999, and there were still funds available. So, we decided to invest those funds to develop the Environmental Recovery Plan for the Old Town to fully replace the sewer and water systems. The preliminary project was approved, and we were granted the funds. (Appendix, Interview, Group B)

The municipal administration exerted real pressure, mostly at the regional level, by claiming that the regional government was not paying any attention to the territories located in the north of the island and, especially, the Municipality of Sassari:

> Since the very beginning, the relations with the regional government had been honest but also blunt. We reaffirmed to the regional government the pivotal role of the [Municipality of] Sassari. One of the objectives was to request substantial resources for the Municipality of Sassari, because the city was at the heart of an area that needed to interconnect, collaborate and create synergies to overcome the severe crisis. When the chemical plant shut down, the areas of Sassari and Porto Torres suffered from a major decline in local production. The crisis also affected the housing sector and employment in the building sector; in Sassari, real estates had to be regenerated because sales had dropped; the city had expanded on a vast territory, and some areas were too far from the city centre. [Our goal was] to reconnect those areas with the Old Town, to revive it. The Old Town had been abandoned by its citizens and trade was stagnating because all the activities had moved to the industrial outskirts of the city, that is, the area of *Predda Niedda*. We tried to tackle the issue, but we were not successful. (Appendix, Interview, Group A)

Requests were being submitted during meetings that were held at the regional offices or during meetings organised with regional advisor or held during the institutional activities of regional politicians:

> We went to Cagliari, to the [headquarters of the] regional government. We had a good relationship with the regional Councillor, with City councillor for public works Luigi Lotto, and also with the regional Councillor and the regional government as a whole. President of the regional government [Federico] Palomba used to visit Sassari often. On those occasions, we used to organize institutional meetings and public debates, but we also devoted some time to talk about project development, potential funding sources, etc. We used to hold meetings with regional advisors before submitting our requests to the regional Council. In a nutshell, we tried to "pull together" Sassari's ruling class and draw resources to the city. (Appendix, Interview, Group C)

> At that time, Luigi Lotto was serving as a City councillor. He was an excellent councillor, [. . .] who worked hard to obtain funds for the regeneration of the Old Town of Sassari. I think he believed it [was the right choice]. So, he served in the regional Council and supported some projects operating at the Municipality level for some time, such as the reconstruction of the water and sewer systems. He managed to obtain funds for them. The projects were rather outdated and technically poor; however, he managed to obtain major funds by lobbying for them. To be honest, I do not know if it was part of Mayor Sanna's agenda or if it was by chance. Those projects were ready. He went to the regional government, submitted them, explained that he had a plan and that he needed the resources. The regional government had EU funds available in the framework of the 1994–1999 CSF and, being behind schedule, it did not have much time to allocate them. So, it was a win-win situation for both Institutions, and it was easy to be awarded the funds. (Appendix, Interview, Group D)

The municipal administration submitted the request to the regional government but also the national government, using a direct communication and relationship channel with the Ministry of the Environment:

> I almost managed it on my own, being a Director of the Department of public works appointed by the mayor, and the Department of the Ministry of the environment responsible for remediation and water treatment. [. . .] We went to Rome to present several projects and show that there was proper planning and we just needed the resources. It was a gamble. I presented the projects developed in the past by the previous administrations, focusing on city development. But there was one project on the regeneration of the Old Town, [drafted by] three designers: Olmeo, Ruju and Fraghì. The project had been funded by the regional government ten years before. The project was very 'dear' to the regional government and the Ministry. (Appendix, Interview, Group D)

The use of that channel by the municipal administration resulted in a kind of 'institutional short circuit' since, during that phase, the Region was sidelined in a decision-making process that only involved the Italian Ministry of the Environment and the Municipality of Sassari:

> At that time, there was a little tension between the Ministry [of the environment], the regional Council and the Municipality of Sassari, because substantial financial resources had been assigned to [the Municipality of] Sassari without any involvement of the regional government. (Appendix, Interview, Group A)

Under the 1994–1999 CSF, the Italian Ministry of the Environment played a leading role, an active role of coordination and control in the management of funds, downgrading the Region of Sardinia to a secondary role:

> The Ministry [of the environment] was one of the stakeholders; we talked with the Ministry, that had the responsibility to check all the opinions; we also received a table with feedback

> on the project. It was a standard table that was used for all municipalities, that included the opinions of the archaeological, architectural, civil engineering *Soprintendenzas*, for seismic, hydro-geological and military constraints. (Appendix, Interview, Group A)

The regional administration was only responsible for the management of the funds, which was divided into three phases: the planning, the transfer of funds and the final reporting:

> [The Region] focused on the schedules to make sure that [projects] would be able to spend EU funds by the set deadlines [...]. At one point, the Region had a meeting with the managers of the Municipality, and they started blaming each other; some representatives of the regional government claimed that they would never be able to meet the deadlines. They argued. I was responsible for the procedure, and I was asked to have a meeting also with Service director Marco Melis. They asked me if the project could be completed on time, and I replied that I thought so. I was also asked why I had not started the competitive bidding procedure, and I replied that I had submitted the project to the *Soprintendenza* for an opinion. (Appendix, Interview, Group D)

> The Region did not seem to interfere much because, once we were granted the funds, we talked about deadlines, the recognition of certain costs, the awarding of contracts [...]. We did what we had written; we talked with the Region about bid price discounts and fund maturity. However, I had the impression that they were sincere. Once EU fund are transferred, you can just start your project right away. (Appendix, Interview, Group C)

Under the following 2000–2006 EU funding scheme, national ministries lost their major role in the process of coordination and control, and they were cut off from the decision-making process. All functions were taken on by the Region that remained the only contact point between the municipalities and the EU:

> In the former CSF, the Ministry of the environment had an active role because it could ask for progress reports, [technical] opinions, information about the bidding procedures. Afterwards [during the following Framework Programme], things changed. The Ministry transferred funds to the regional government, which in turn transferred funds to us. Under the 2000–2006 Framework Programme, the Region played a leading role at all stages: from the allocation of funds to the communication and coordination of the Municipality and other bodies. The Ministry was just cut off. (Appendix, Interview, Group B)

> The regional Planning Centre was the control room that verified the compliance with the terms for the expenditure of EU funds. [...] There was an around-the-clock dialogue with the regional Planning Centre, because the Centre monitored how funds were being spent; therefore, it was a continuous collaboration. I personally went to Cagliari several times, to the regional Planning Centre, during those years to monitor and deliver the progress reports and to discuss critical issues. But especially to submit the financial reports. (Appendix, Interview, Group A)

In addition to the EU, the Ministry of the Environment, the Region of Sardinia and the Municipality of Sassari, the dominant coalition also consists of two other players: the *Soprintendenza* of archaeological heritage of the Sardinia Region and the *Soprintendenza* of monumental heritage of the Provinces of Sassari, Olbia-Tempio and Nuoro. They have the legal resources but also specific expertise in archaeological protection and monumental protection, respectively. Regarding the ability to steer the decisions made by the *Soprintendenzas*, some preliminary considerations are worth mentioning. As for the *Soprintendenza* of archaeological heritage, it has been provided with legal resources by the legislative body only through the introduction of Legislative Decree 163/2006, in particular Art. 95, which provide for the contractor's obligation, if public works are contracted, to request the competent *Soprintendenza* for the preliminary project approval. This obligation was introduced in 2006 when the legislative decree came into force. Before that date, the request was optional, and works could be suspended only if archaeological remains were found. Therefore, during that period, the collaboration between the institutions was vital:

> Before the years 2005/2006, when the preliminary archaeological investigations were introduced by the law in the framework of public works, they all depended on the collaboration between different institutions. Public authorities had no obligation to include any cost for the archaeological surveys in the budget. Those costs were introduced later. So, for instance, in Sassari, since the risk of finding archaeological remains beneath the city was high, the *Soprintendenza* of archaeological heritage prompted preliminary discussions. When the Municipality forwarded us the first preliminary project, we organized meetings to explain the requirements and the common goal in the field of prevention rather than improvement. At that time, the *Soprintendenza* could not require any archaeological assistance and preliminary investigations to be paid by the project. However, it could suspend work if buildings or layers of archaeological interest were found during excavations. So, we 'diplomatically' talked to the technicians of the Municipality and explained how a preliminary project could be designed, before works started, to manage the archaeological risk. We wanted to prevent any damage to the archaeological heritage and the Municipality wanted to prevent the interruption of works and any pointless issues, by dealing with them in advance. (Appendix, Interview, Group D)

Despite the lack of a specific set of laws during that period, a written agreement was drafted as a result of a series of meetings among the representatives of all the players, governing the entire policy-making process and the relations among the players:

> An agreement was signed by the Municipality, and the *Soprintendenza* of archaeological heritage, providing that preliminary investigations and works should be aided by an archaeologist. We could not carry out those tasks directly because we had already too much work to do, but an external archaeologist could be paid by the Municipality. [. . .] We agreed that an agreement was necessary, providing the instructions on follow-up procedures, meetings with the Municipality, and the appointment of an external archaeologist. The Municipality would take care of payments. Based on this agreement, the work carried out in the Old Town of Sassari enjoyed the support of one or more archaeologists. (Appendix, Interview, Group D)

Additional legal resources were made available to the *Soprintendenza* of monumental heritage in 2006, when the Regional Landscape Plan came into force, and the Code of Cultural Heritage and the Landscape was amended. In the past, the role of the *Soprintendenza* had been limited to merely revoke landscape permits already granted by the Region or the Municipalities. According to Art. 146 of the Code, that power was greatly extended, providing that the Region and the Municipalities delegated by it sent to the *Soprintendenza* the documents submitted to obtain the authorisation, accompanied by an explanatory memo and a draft proposal for the measure. The *Soprintendenza* has the responsibility to provide a mandatory and binding expert opinion. The balance of power among players was not determined by the possession of legal resources but by the representatives' mediation skills:

> Since the beginning of the Sassari Old Town regeneration project, the Municipality and our Office have been collaborating closely. In the beginning, there was a little tension because the *Soprintendenza* had submitted to the Municipality requests, which had been regarded as strict and severe. I must say that, over time, we had nice cooperation. Even the Municipality admitted that those requests, despite being perceived as too strict, resulted in a greater work quality. [...] During the first round of works, our scope of action was limited to the protection of monuments because the landscape plan had not yet entered into force and, therefore, Old Towns were not subject to landscape protection. We were responsible for authorizing work on the structures found underground during the excavations since we had to authorize their partial or full relocation to make way to pipes. (Appendix, Interview, Group D)

The relations between the two *Soprintendenza*s (the *Soprintendenza* of archaeological heritage and the *Soprintendenza* of monumental heritage) were not governed by rules, codes of conduct or agreements, but stemmed from a strong sense of cooperation that, over time, had become usual, creating an unwritten set of rules of conduct that was being systematically followed:

> A close informal collaboration between the *Soprintendenza* of archaeological heritage and us started. Since the works were carried out in the Old Town and the layers and findings gradually recovered during the works dated back to the medieval or post-medieval periods, we were responsible for their excavation, while the other *Soprintendenza* was responsible for their protection. (Appendix, Interview, Group D)

The relations among the players who had a prominent role in the execution of works were regulated through the establishment of a committee which met weekly and service conferences consisting of decision-makers who had different interests but pursued a common objective, namely, to ensure that the execution of works was performed efficiently and effectively:

> There was a permanent consultation with the stakeholders involved; an internal committee was summoned including several Departments and representatives from *Confcommercio* [the Union of local enterprises]; the committee coordinated works in the Old Town, but also around thirty additional road construction sites, and under that committee the new roads of access to the city were built. (Appendix, Interview, Group C)

> The instrument [called] *Service conferences*, that is, the consultation process of bodies concerned in a specific procedure, has become very popular over time, in particular during the Old Town regeneration process. Those consultations played an important role because they gathered all the decision-makers, requesting their expert opinion on specific matters. [...] The consultation process was useful to solve several time-consuming issues and complex bureaucratic procedures. [...] Dialogue was the most efficient instrument to make a decision. (Appendix, Interview, Group A)

The main function of the committee and the service conference was to overcome the bureaucratic stalemate that was caused by the need for different types of authorisations that were to be released by different bodies, the latter being independent of each other and having exclusive jurisdiction. Using this tool, we had a direct dialogue on a number of issues that allowed an agreement between stakeholders and, at the same time, generated a decision that was binding for all:

> To secure work continuity, avoid the submission of written requests for every single action, and formalize all the authorization requests to the Municipality, [...] a steering Committee was summoned, including the representatives of all the institutions involved: our *Soprintendenza* of monumental heritage and the *Soprintendenza* of archaeological heritage, the representatives of the public and private companies governing sub-services, such as Telecom, Medea, [...] Enel, [...], the representatives of other institutions, the civil engineering department, other municipal Departments. We used to meet on Thursday afternoon, at least for an hour; usually, the meetings lasted much longer, and we used to make decisions from time to time as worked progressed. We did so to avoid suspending works. (Appendix, Interview, Group B)

> The decisions made at that meeting were executive, that is, we no longer needed any additional step to approve our decisions. Decisions could be executed immediately because the Committee included the highest representatives appointed by the *Soprintendenza* of archaeological heritage, Enel, the procurement agent, the contractor, Medea, etc. That is, all the decision-makers involved in the project. It was a great success. At times, the meetings were attended by the representatives of other companies, traffic police officers and colleagues from the Municipality who were responsible for institutional communication. The Committee provided its support until the testing phase. (Appendix, Interview, Group A)

Degree of Congruence of Cognitive Maps Within the Dominant Coalition

The second dimension examined in terms of the institutionalisation of the governance system relates to the (b) degree of congruence of cognitive maps within the dominant coalition.

The congruence of cognitive maps among the actors of the municipal administration was improved by the Municipality of Sassari since the mid-1990s to better exploit the opportunities generated by EU funds. Such awareness led to a reorganisation of the Municipality:

> Since the mid-1990's, the municipal Department of public works was reorganized: new managers were appointed to take care of public works or maintenance, or urban planning, and also new heads of office were appointed with well-defined responsibilities; the municipal Department of public works operated independently from the others, without the coordination of a top manager, except for the Municipality's General director. During those years, the Technical Department was also changed. (Appendix, Interview, Group B)

> The reorganization also affected the programming sector, in particular for the identification of and application to regional funding calls. (Appendix, Interview, Group D)

The reorganisation was followed by the recruitment of new administrative staff specifically for the public works department, mainly involved in the implementation of the works requested by the political bodies of the Municipality of Sassari:

> From an administrative point of view, it was quite a good time since several staff members changed at the Municipality. At that time [...], several technicians and officers were hired to

take care of the design and implementation of public works. An office was created to manage those works, but also colleagues for other offices supported their work beautifully. The administration was aware of this; the entire administration was aware of this effort and somehow both the politicians and managers played proactive roles. (Appendix, Interview, Group D)

We succeeded thanks to them, because we were young in 1997, having been trained by Formez and well educated, usually graduated people. (Appendix, Interview, Group D)

Thanks to the organisational and structural reform during that period, the political body created an efficient administration, which had a considerable workload since he had to perform a vast amount of activities and meet deadlines; nonetheless, it was able to adequately support the decisions taken by the political bodies of the Municipality of Sassari:

I remember that [Lotto] came and told us that the funds were being transferred. We had to spend them immediately, so we had to publish new calls at once [...] It meant we would be managing thirty to forty projects; it meant having thirty to forty designers and foremen to hire; thirty to forty companies to select; and then all the necessary accounting and reporting. So, we published one-step calls and awarded contracts very quickly. [...] Considering the results, we met all the deadlines, we complete projects and we spent all the budget available. So that was, in my opinion, a remarkable result. (Appendix, Interview, Group D)

We worked very well, but none of us had been trained specifically for it; it was thrilling for every one of us because no one believed that we had done that work in two years, and we had missed the deadline by an inch only in one project. [...] There was general coordination by the chief of staff. The greatest difficulties concerned the authorizations for the use of public spaces; however, everything was concerted, requested, discussed, and clarified in advance; from this point of view, it was a perfect model. (Appendix, Interview, Group D)

I remember that the Integrated Territorial Project for the Old Town required €25 million for the design phase; it took us 6 to 8 months only to award the contract through an EU call procedure. For that reason, the Municipality chose to split the tender into four functional lots for a total amount of €25 million. All the rules were strictly complied with. It was a strategic goal for the Municipality, so there were no problems. (Appendix, Interview, Group D)

Tribute must be paid to the technical and administrative staff of the Municipality, both the executives and the managers who helped me successfully submit a flawless project to the regional government. The project was selected by the Region first and the EU later. Such a good job! (Appendix, Interview, Group A)

The interests protected by the two *Soprintendenzas* of archaeological heritage and monumental heritage were pivotal to the prompt delivery of a preliminary study amending the project, which changed from a merely infrastructural design to a historical protection and restoration initiative. Therefore, right from the beginning of the design phase, the *Soprintendenzas* requested that specific steps be followed during the execution of the works, and the Municipality did so:

A preliminary study was approved by the *Soprintendenzas* of monumental and archaeological heritage, and it was then amended. We introduced substantial changes to the project and, for the first time, new processes were added, namely the preventive demolition of the tarmac layer, the recovery of the original paving, the graphic and photographic survey with the prior numbering of all paving stones, the cataloguing of all the pebbles. [...] We collaborated very well with the *Soprintendenzas*. (Appendix, Interview, Group D)

While there was a congruence of the cognitive maps between decision-makers and administrative staff within the Municipality and between municipal staff and the *Soprintendenzas*, the relationship between the Municipality and the regional government was quite different, resulting in a mismatch of the cognitive maps within the two institutions. This is mainly due to a change of the rules of the game at the political level and to uncertainty and failure to share practices and rules at the bureaucratic and administrative level:

> The only problems we had with the Region regarded the use of EU funds. As a matter of fact, the Region had decided to use those funds in a certain way, but in the end, it changed its mind and allocated funds differently. The Municipality of Sassari was negatively affected by that decision, having made a consistent choice with a technical staff that could carry out project very well. Indeed, our calls were well designed and matched the model perfectly. [At the Municipality], a great work was completed to select all the projects eligible for those calls. And the selection was performed by assessing the impact, the extent, the characteristics of the project, the involvement of staff, the costs. The Region decided instead not to follow that approach and applied other criteria. That decision had a negative effect on the Municipality [of Sassari]. (Appendix, Interview, Group B)
>
> The transfer of competencies from one Ministry to another, or, within the Region, from one *Soprintendenza* to another, or even from one head of office to another, caused problems; for example, we were asked to provide the same documents multiple times. [. . .] There was not a single procedure in the Region with regard to EU funds. Each *Soprintendenza* does as it pleases, and you never know what to do. (Appendix, Interview, Group D)

Degree of Structuring and Impermeability of Gatekeeping

The third dimension in the institutionalisation of the governance system relates to the (c) degree of structuring and impermeability of gatekeeping, particularly concerning the actors who are able to control, at least at the major nodal points, the political, legal and economic–financial resources of the decision-making process.

With regard to Sassari, the key decision-making steps throughout the policy process and, therefore, the level of structuring and gatekeeping impermeability ebbed and flowed. Initially, the redevelopment of the historical centre was deemed necessary due to the need to rehabilitate the water sanitation system, considering that the area had an outdated infrastructure network. This and the opportunity to have access to EU funds gave the administration access to major financial resources. Thanks to the approval of the redevelopment projects submitted, a whole range of relations among players developed concerning the management of financial resources:

> We had relations only with the Region and the Ministry of the environment, and I guess there was an overall plan because all the activities primarily concerned the rehabilitation of the water-sanitation system of the Old Town. Infrastructural works were funded first to renovate the sewer and water supply system of the Old Town. After starting a dialogue with the *Soprintendenza* of archaeological heritage and the *Soprintendenza* of monumental heritage, the projects widened their scope to the full restoration of the original paving. But first, they renovated the water-sanitation system. (Appendix, Interview, Group B)
>
> The ROP Sardegna 2000/2006 Programme gave the Municipality of Sassari the opportunity to upgrade the Old Town infrastructure, to renovate the water system, the sewage

system, and later on, the electrical, landline and gas distribution network. Taking advantage of Measure 5.1 of the ROP Sardegna 2000/2006 Programme, the Municipality of Sassari was awarded a €25 million loan funded in the framework of the PITSS01 call, that is, the Integrated Territorial Projects of the Municipality of Sassari. [. . .] If I recall correctly, it was one of the most comprehensive plans funded by the ROP Sardegna 2000/2006 Programme. €25 million was a remarkable allocation. (Appendix, Interview, Group C)

In addition to the resources funded by the EU, the administration had to cofinance the works. As it was usually the case, it cofinanced part of the loan:

It was co-financing. The Municipality had to provide part of the funding. Indeed, EU funding is granted only if co-financing is provided by the Municipality. (Appendix, Interview, Group B)

During the initial phase of the policy-making process, economic and financial resources were managed and controlled by the Municipality and the Region. The municipal administration managed funds, while the Region was responsible for the audit procedures. The direct access to resources, therefore, was allowed only to the players who exercised such functions:

I must say that the relations with the regional government were extremely good. EU funds were acquired by applying to specific regional calls. Afterwards, the Region no longer managed the activities nor implemented the works. (Appendix, Interview, Group A)

All communications regarding the amendments and the variations to the projects were forwarded to the regional government. The Region had full control of spending and financing, since we could not make decisions regarding bid price discounts and we had to ask for the regional government's authorization. (Appendix, Interview, Group C)

Although the economic and financial resources, in fact, were managed by the Municipality and the Region, the two *Soprintendenzas* of archaeological heritage and monumental heritage heavily affected project sustainability through their legal resources, since they forced the City Council to hire archaeologists (selected by the *Soprintendenzas*) to work directly on the field, in order to avoid further delays in the execution of works:

Through an agreement, the *Soprintendenza* of archaeological heritage and the *Soprintendenza* of monumental heritage provided their archaeologists, who monitored the works and were paid by us. In any case, each *Soprintendenza* could suspend works, makes all the necessary checks, resume or stop works, order to remove something or request to pay extra attention. Then, to avoid delays, we solved the problem by bringing departmental archaeologists in the field and paying them with our funds. (Appendix, Interview, Group A)

The *Soprintendenzas* choose the archaeologists, and they were paid using the funds allocated for the Old Town; from that point of view, there was excellent cooperation, in my opinion. As things were done, the *Soprintendenzas* served as observers and supervisors at the same time. (Appendix, Interview, Group B)

Although the two *Soprintendenzas* of archaeological heritage and monumental heritage did not exercise direct control over economic and financial resources, their intervention through the use of their legal resources resulted in an indirect control over them, leading to a review of the preliminary draft project, due to the additional expenses, which initially had not been included in the budget:

> Those projects had to pass several evaluation steps. The regional technical Committee had to verify the project technical and administrative quality. The *Soprintendenza* of archaeological heritage and the *Soprintendenza* of monumental heritage took care of the second evaluation. Since works were carried out within historical and archaeological protected areas, the projects were submitted to the *Soprintendenzas* for their opinion. An extensive dialogue started with the *Soprintendenzas*, which led to amendments because the project had to change from a purely infrastructural project to a protection and regeneration project. [...] This work included archaeological assistance during construction work. There was a major investment in archaeological surveys because the archaeologists were paid from project funds. But great discoveries were made when the roads of the 'old Sassari' were found, i.e. dating back to the 10th, 11th century B.C. Forgotten areas were found. So, we did a great job! (Appendix, Interview, Group B)

Technical and organisational resources, especially about the design stage, had first been managed only by the Municipality, which had performed that function regardless of sector-specific and micro-sectoral interests, overseeing access to the dominant coalition under a predominantly impermeable system:

> This is the idea: [if] your design takes into account all instances, the design is participatory, but the approach may cause delays. We did not implement a participatory approach. The project was developed by the Municipality only: for example, a shop owner could be involved to decide how the main square, *Piazza d'Italia*, had to be arranged, also because the Municipality had full technical discretion. [...] Third parties were not involved at any stage of the decision-making process. (Appendix, Interview, Group A)

The differences between the players of the dominant coalition, and specifically the Municipality of Sassari and the *Soprintendenza* of monumental heritage, on the redevelopment works, led to a revision of the initial project by the municipal administration, which, being unable to review the initial design solutions due to the tight schedule (at the risk of losing funding), decided to work on another area of the city; specifically, it decided to redevelop *Piazza d'Italia*, instead of redeveloping the lower part of *Corso Vittorio Emanuele*. During that phase, the gatekeeping level was much lower than the previous phase, as the dominant coalition made decisions based on the influence of external pressure groups, which affected the decision and determined its reformulation:

> The reorganization of the lower [...] part was quite impossible. In the beginning, there had already been several 'noes' from the *Soprintendenza* of archaeological heritage and the *Soprintendenza* of monumental heritage, so we had to redesign everything. We were unsuccessful and then we decided to focus on *Piazza d'Italia* and reorganize the area, that was the very heart of the city centre. (Appendix, Interview, Group D)
>
> The first idea for *Piazza d'Italia* was an underground car park, trying to minimize its impact. Underground car parks were popular at that time. The *Piazza Fiume* underground car park was being completed, others were already being planned, and we thought we could make one under *Piazza d'Italia*. The project was described in a draft proposal, but the citizens basically rejected it. There was a major opposition because it was seen as a scar on the original square. So, we decided it was not practicable. (Appendix, Interview, Group B)

In addition, during the implementation phase of the regeneration policy, the level of gatekeeping decreased due to the pressure exerted by the retail business owners, who managed to be granted an amendment to redevelop some areas of the Old Town:

The Municipality changed its mind regarding *Via Rosello*, a commercial road in the city centre. [. . .] We organized several meetings with retail business owners, and at one point, we decided to amend the project by excluding *Via Rosello* from the implementation plan. (Appendix, Interview, Group C)

An intervention had been planned in *Via Rosello*, but we postponed it because we believed that the redeveloped area had been affected too much by the delays of the *Piazza Castello* construction site, and we preferred to postpone *Via Rosello*, being another commercial axis within the Old Town. (Appendix, Interview, Group B)

The Municipality of Sassari also launched a number of distributive tax policies to relieve the pressure on the dominant coalition and, therefore, on the decision-making process, trying to increase the level of sharing of the action taken through the use of such instruments:

We thought about businessmen because shops had suffered too much; also, there were also tax concessions to support them. (Appendix, Interview, Group C)

Regardless of other economic policies, we promoted local tax relief or the allocation of funds for investments on businesses located in the Old Town; both us and the Ganau Council did this; so, we acted through the fiscal and financial incentives levers. (Appendix, Interview, Group A)

3.4.1.2 The Quality of the Urban Regeneration Policy

Degree of Responsiveness

Turning to the analysis of the quality of the urban regeneration policy, the first dimension examined is the (a) degree of responsiveness, which can be assessed by comparing the expectations and objectives of the intervention shared by citizens as well as the local ruling political class, with the actual results achieved by the policy.

In the case study of Sassari, the degree of policy responsiveness can be evaluated by comparing the expectations and the goals of the actions taken, shared by the citizens and the ruling class, i.e. promoting regeneration of the Old Town through the establishment of new businesses and an increase in the resident population, with the results achieved. The interviewees involved provide a useful perspective to understand the situation fully. However, a preliminary consideration is required: the works performed consisted in the rehabilitation of infrastructures only, except for the most important public buildings, meaning that the investment paved the way for future development:

Perhaps from an architectural point of view, we could have improved something more, but the goal of the project was achieved: paving the way for the regeneration of the Old Town. The adjustment of public squares and the installation of new underground systems were the first steps to take to set the urban regeneration process in motion. (Appendix, Interview, Group A)

[Sassari's] Old Town had interesting and unique architecture. [. . .] We decided to put all our efforts into the Old Town. We managed to regenerate and redevelop the city centre, equipping it with modern systems and, at the same time, safeguarding the values of solidarity, coexistence, common roots, habits. It was very beneficial for the city, and it fostered a better quality of life and coexistence, at a time when, around the globe, the outskirts of cities become unspecified locations, dormitories, and often 'non-places.' (Appendix, Interview, Group B)

Despite the actions taken, the City Council failed to generate an opposite trend that would result in a regeneration of the Old Town through the involvement of private operators. In addition to the total lack of interest of local shop owners in investing in the area, the depopulation of the Old Town continued, and traditional citizens moved away and were replaced by people forced to live in poverty and hardship conditions:

Unfortunately, the regeneration of the Old Town was seen by local policy-makers as a city planning and building operation. [...] The social aspect was not taken into account at all. In the city centre, residents were not even considered. In the city centre, several works had been carried out, and buildings had been built. That is true. Just think that 90% of the people living in the lower part of the Old Town are migrants. (Appendix, Interview, Group E)

The middle class has left the Old Town and moved to the [suburbs]. Therefore, the area is degraded; residents are mostly poor people. Rents [are] cheap, so low-income families move there. [...] Landlords have exploited migrants. Degradation is visible. [...] There is no future for the Old Town if it is not regenerated. Unfortunately, the policies focusing on the Old Town have been unsuccessful; [policy-makers] have been more interested in developing the outskirts of the city because investments in construction sites have moved there. (Appendix, Interview, Group E)

The Old Town no longer belongs to the people of Sassari. [...] We have built sky-scrapers, we have depleted the local community, especially the Old Town. Nowadays, modern techniques can help plan and redevelop an Old Town by upgrading existing properties; for instance, old buildings might be hard to live in by older adults. Just think about the old flights of stairs. While preserving the outside of the buildings, much can be done to upgrade the inside: for example, by installing elevators, changing some features to make life easier. The people of Sassari can no longer live in those conditions. Several buildings were built using tuff, which creates an intolerable degree of humidity, but new building techniques can be used to renovate them. (Appendix, Interview, Group D)

Unfortunately, the Old Town is only slightly affected by the regeneration process: only the areas of *Piazza Tola* and *Via Pettenadu*. Down from *Via La Marmora*, degradation, poor public lighting, smelly underground systems, and high humidity levels are widely common. *Via Rosello* had been the main shopping area in the Old Town until the Sixties. Today it is a ghostly, closed, dark, humid street, with one open shop every five closed. Moving towards the *San Donato* district, the degradation is even greater, and shops are limited to small businesses run by migrants. So, no one can argue that, from *Piazza Tola* down, there has been any kind of regeneration, from any point of view. No social, architectural, structural or trade recovery was attained. (Appendix, Interview, Group C)

Where do people go now? To the malls. They are open seven days a week. Yes, maybe something could have been done looking twenty years ahead. [...] So people go there for everything, they walk, they get ice cream, they go to the movies: that is the city centre now. [...] Now, our Old Town, down from the *Quattro Colonne* theatre, is only inhabited by migrants. To give a new life to an Old Town, you have to bring life there. But what if all the people go to the mall? (Appendix, Interview, Group C)

Degree of Cooperation

As far as the second dimension of the quality of the urban regeneration policy is concerned—that is, the (b) degree of cooperation—at first, the policy was characterised by a low tendency of the players who interact within the policy network to cooperate. During that first phase, the Region played a minor role compared to other stakeholders, because between the Municipality of Sassari and the competent Ministry, there was a 'privileged relationship' that, as a matter of fact, excluded the Region from the decision-making process, only granting it a minor role. Relations within the network partially improved later when, starting from the planning for the 2000–2006 structural funding period, the Ministry left the dominant coalition, and all its functions were acquired by the Region, which started directing the whole process.

> [Our relations with the] regional government were not good at all: on the one hand, the regional government was very late in the planning phase. In the 1994–1999 CSF framework, we received the funds only in 1997 [. . .]. I started working in May 1997; the calls were published in 1998–1999. The funds had to be reported and fully spent by 2001. So, during that year and a half, we went crazy. [. . .] The regional government only had to transfer funds and regulate the procedures, but the Ministry was the second and final controller. For example, the regional government discussed final decisions with the Ministry. [. . .] The regional government had procrastinated for several years and, eventually, blamed the municipalities, requiring them to hurry up, otherwise, they would have revoked the funding. So, we did hurry up. (Appendix, Interview, Group D)

A second cause was the conflict between the players who interacted within the decision-making network. The main cause was that the interests pursued by the players, within the coalition, in some stages of the process, were clearly in contrast to each other, which resulted in opposing clashes, a certain instability and little cohesion of the dominant coalition. Those opposing clashes also led to the redefinition of policies, resulting in reprogramming and the transfer of works in other areas of the city:

> As an auditor, I attended a meeting of the city planning Committee of the regional Council where several councillors from Cagliari were complaining that the Ministry was to blame. Since then, the following funding schemes have been managed again by the regional government, not by the government. (Appendix, Interview, Group D)

Degree of Functional Congruence

As far as the third dimension of the quality of the urban regeneration policy is concerned, that is, the (c) degree of functional congruence, with particular regard to the capacity/incapacity to calibrate and steer programmed actions towards specific social target groups, we can turn our attention to two different categories of actors towards whom the environmental protection policy was directed.

In the case of Sassari, we can say that the regeneration process has not fully come to fruition yet, but only partially since the intervention of private players was found to be virtually absent. The effects are the following: (a) the lack of interest by private stakeholders to invest in real estate in the city centre, (b) the business owners' low take-up of regulatory and financial instruments provided by the Region to support the regeneration of the Old Towns and (c) the presence of strict regulatory constraints for architectural protection. These aspects have been particularly highlighted by the players and the privileged witnesses as part of the interviews:

> There was no regeneration of houses and blocks. The Old Town was abandoned by the people of Sassari. [Although] all the works for the regeneration of underground systems were completed, houses were not restored, so the degradation remained unchanged. The project restored some blocks of buildings, but nothing else was done, except for some minor works by private investors. (Appendix, Interview, Group E)
>
> I believe that giving oxygen to breathe, giving air to the Old Town is really important to make it a habitable space. It makes no sense to keep run-down old buildings because they have no value whatsoever, and they are not habitable. Run-down building cause neighbours to move away because they mean humidity, rats, dirt. Protecting the Old Town is important not just because it is "old" and the "ugly". (Appendix, Interview, Group C)
>
> To regenerate the Old Town, you also need to renovate buildings, and it is unlikely that an owner decides to invest resources to renovate his or her own property, it the degradation of that area is so huge. Renovating a single building or a small block of buildings, in the end, is useless and gives no benefit. (Appendix, Interview, Group E)
>
> In my opinion, public investment was impaired by the lack of private investment, that is, the opportunities offered by the law for Old Towns were not seized. I strongly believe that to regenerate an Old Town or urban spaces, the coexistence and participation of public institutions is necessary. Public institutions can give momentum, they can provide the guidelines, but then the process must also be embraced by private investors. [...] Integrated approaches can promote Old Towns. (Appendix, Interview, Group E)
>
> The Municipality of Sassari did what it had to do, that is, regenerate infrastructures. Unfortunately, we continue to see a very degraded Old Town, but that depends on the owners, the residents, and other factors that go beyond the will of the Municipality. (Appendix, Interview, Group A)
>
> In my opinion, more regeneration actions are needed to promote the redevelopment of the buildings. We must encourage the regeneration of private buildings, and important decisions must be made to remove useless spaces occupied by awful and impracticable hovels. [...] We also need to launch incentive policies. (Appendix, Interview, Group C)
>
> An Old Town and the surrounding area is alive and vital if it complies with the IAA rule: being Inhabited, being Accessible and being Attractive. In the Old Town of Sassari, the living-residential function has been mortified. Some have confused the "precarious" with the "old and historical": on the one hand, we have precious historical buildings; on the other hand, we have crumbling buildings. So, the residential function must be guaranteed by making a distinction between historical buildings that must be preserved by all means because they testify the history of the city, and what is crumbling and precarious, that should be torn down. Accessibility: in the context of city planning, a mobility plan is an essential element because in critical areas, such as the Old Towns, accessibility must be guaranteed. So, the important thing is not to make the area accessible by car, but to make it easy for people to get there; this means developing a complex system of car parks, suitable public transports, wide pedestrian areas. Attractiveness: the attractiveness is linked to the functions that the centre can provide. Otherwise, it is only a dormitory. The appeal is therefore associated with the existing services, which also require consistent urban planning decisions. So, that is why in Sassari, the redevelopment of the Old Town was not consistently planned, unlike other areas. (Appendix, Interview, Group B)

Despite the lack of proactiveness of private investors, examples of the regeneration of the Old Town were reported in a few areas, showing that results are satisfactory when public and private investors collaborate:

> There are areas where that recovery has taken (and is taking) place, such as *Piazza Tola*, where there has been good cooperation with shop owners. Where the 'new' was best used, good results are visible. Let's think about the *Corso*, the highest part of the *Corso*, along with *Piazza Tola*. Culturally speaking, that area has become a focal point of the city centre. (Appendix, Interview, Group C)
>
> Today, we can say that *Piazza Azuni* and *Piazza Tola* are two extraordinary examples. In particular, *Piazza Tola* has really developed: the old street market is no longer there, but it has become a place for social gatherings, especially for young people. (Appendix, Interview, Group B)
>
> Today, *Piazza Tola* has been rediscovered by the citizens of Sassari, and it makes me very happy to see people gathering there. (Appendix, Interview, Group A)

3.4.2 *The Typical Case Study of Alghero (1995–2006)*

Alghero is the second most populated city of northwest of Sardinia and the fifth most populated city of the island (Fig. 3.3), with 43,931 inhabitants (ISTAT 2019).

3.4.2.1 The Institutionalisation of the Governance System

Degree of Functional Specialisation and Interdependence Maps Among the Players of the Dominant Coalition

The first dimension examined in terms of the institutionalisation of the governance system relates to the (a) degree of functional specialisation and interdependence maps among the actors of the dominant coalition.

The Municipality of Alghero, as well as the Municipality of Sassari, began planning the regeneration of the Old Town in the mid-1990s. The Municipality of Alghero began to design a series of projects which were first funded by loans only. During that period, as a result, the dominant coalition of the policy network mainly consisted of three players, that is, the Municipality and the two *Soprintendenzas* of archaeological and monumental heritage:

> In the Nineties, the agenda of the Municipality of Alghero included the regeneration of the Old Town. The agenda was consistent over time, mainly because the same political coalition ruled over the city between 1994 and 1998. The electoral agenda included urban renovation, environmental protection, and the conservation of historical and cultural heritage. (Appendix, Interview, Group A)

The enlargement of the dominant coalition through the inclusion of the Region of Sardinia only occurred in the late 1990s:

> The Municipality of Alghero has received ROP funds for the renovation of the promenade along the ramparts, in the framework of a project called "Muralla 2000". The funds were

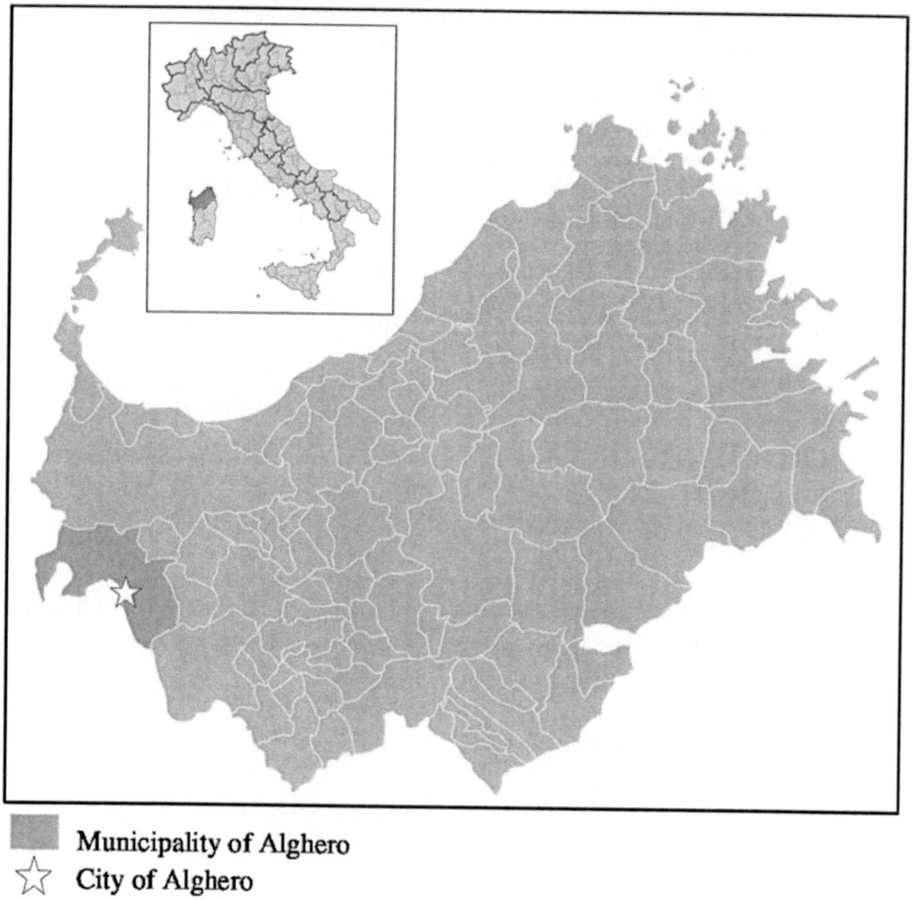

Fig. 3.3 The municipality and the city of Alghero

> transferred probably in 1999. It was direct funding form the regional government that had the right to assign resources directly. This right was revoked later on when the procedures changed. (Appendix, Interview, Group B)

Unlike Sassari, the Ministry of Public Works was not present in that dominant coalition of the policy network; all the governance functions were exercised by the Region:

> They were ROP funds managed by the regional government. In other cases, Urban Regeneration Programmes had been implemented by the Ministry of public works. That was not Alghero's case. (Appendix, Interview, Group B)

Unlike those between the Region of Sardinia and the Municipality of Sassari, the relationships between the Municipality of Alghero and the Region have always been marked, right from the beginning, by a proactive collaboration. There has never been any 'institutional friction' and communication, and the reporting system was structured and well-defined, sometimes even through the signing of memoranda of understanding which clearly described the tasks of each political and administrative body:

> The main local stakeholders were the Municipality, the coastlines, and the industrial areas which were to be redeveloped. The regional government took care of the paperwork for design verification and its compliance with the goals of the EU funding scheme. The interaction [...] is typically Italian: very bureaucratic, often very complicated. [...] Especially for this type of project financing, the regional government focused on the financial aspects, project compliance and its timely implementation. (Appendix, Interview, Group C)
>
> When we were ready to go, by approving and contracting works, [...] a memorandum of understanding was signed. [...] The agreement regulated the relationships between the Region and us, so the document included all the deadlines and clauses, such as the constraints of use for the car park, that was required to be used as such for the following 10–15 years. (Appendix, Interview, Group A)

The two *Soprintendenzas* of archaeological and monumental heritage and the Municipality of Alghero have acquired legal resources over time, each of them in its own sphere of competence. During the period before the introduction of the rules which defined more precisely the legislative framework, the relationships among the Municipality of Alghero and the two *Soprintendezas* were characterised by cooperation, mediation and agreement but also at times by disagreement regarding the mode of intervention:

> The whole territory of the Municipality of Alghero is protected, and several buildings, such as the walls, are protected by monumental restriction orders. At that time, urban policies were fostered based on the opportunities offered by the ROP Funds. If the works were planned in a monumental area, we had a fairly extensive role in the decision-making process. (Appendix, Interview, Group D)
>
> I went personally to meet the Superintendent to discuss solutions since Alghero had a very wide area bound by the *Soprintendenza.* So, it was essential to obtain permission from the Superintendent for any work, for any project. (Appendix, Interview, Group B)
>
> It was always an obstacle course because all the works were subject [...] to the constraints of the *Soprintendenzas* and their directions [...]. Both with the Santa Chiara complex, the pavement of the ramparts, and the "lo Quarter" project, [...] the *Soprintendenzas* even decided the type of materials to be used and each and every architectural element. [...] At that time the two institutions directly negotiated the decisions; the officers and the designers were requested to submit the projects directly to them, and the *Soprintendenzas* has the right to request and reject the amendments. (Appendix, Interview, Group B)
>
> Very often, a representative of the *Soprintendenza* of the monumental heritage came to the worksite, and we took decisions directly there, whenever needed. Excavations were made in collaboration with the *Soprintendenza* of the monumental heritage and the *Soprintendenza* of the archaeological heritage, especially within the Old Town; later on, all the relics found underground were protected and always documented. (Appendix, Interview, Group D)
>
> We had some problems with the *Soprintendenza.* [...] Besides, in those years, two or three supervisors changed, and each one of them had a different vision. For example, I remember the problem with the recovery of *Casa Manno*, where we established a museum. [...] First, the project was not good: we had to change it, they had suggested solutions we could not accept. So, we had to negotiate. [...] Yes, it was often tricky, though I have to say that we always sat around a table and found a solution that could [be good] for both of us. (Appendix, Interview, Group D)

The relations among the two *Soprintendenzas* of archaeological and monumental heritage and the Municipality of Alghero were not based on a general agreement; the three institutions collaborated whenever necessary, and in some cases, they were supported operationally by the university:

> There were many different redevelopments affecting different areas of the Old Town, and they were addressed separately from time to time. So, there was no general agreement. [. . .] There was definitely an input from the *Soprintendenza* of the archaeological heritage, that was also finalized in a written document, whereby we had to report to the Municipality of Alghero any work involving underground works in the Old Town, to be carried out within the boundaries of the ancient walls. Those works had to be submitted for authorization to assess any measure or archaeological surveillance to be implemented. The support of the University was also important for the Municipality of Alghero because, since the early surveys in the Old Town, we have collaborated frequently with the University of Sassari, in particular with Professor Marco Milanese, who supervised the work carried out by the archaeologists paid by the Municipality of Alghero. [. . .] The Municipality forwarded the project to us, we assessed it and requested either the assistance of archaeologists or in some cases, preliminary investigations; and then the Municipality of Alghero signed an agreement with the University of Sassari and the archaeologists. (Appendix, Interview, Group D)

Unlike the Municipality of Sassari, the Municipality of Alghero did not create a dedicated body to overcome any obstacle that could arise in the decision and implementation phases of the regeneration project. The issues concerning the authorisation and the execution of the actions taken by each political and administrative player were overcome through a series of unwritten rules that were based on the principle of consultation and sharing:

> Talking about the regeneration of urban areas, although several institutions were involved, we sought maximum coordination because that methodology would secure successful results. We tackled the issues connected to the access to the shore by involving all institutions. (Appendix, Interview, Group B)

Degree of Congruence of Cognitive Maps Within the Dominant Coalition

As far as the institutionalisation of the governance system is concerned, the second dimension to be analysed relates to the (b) degree of congruence of cognitive maps within the dominant coalition.

The typical case study of Alghero shows a high congruence of cognitive maps among decision-makers and the Municipality of Alghero since the latter had changed its organisation to have an effective group of people capable of seizing all the opportunities offered by EU funding schemes:

> At the time, Alghero was as an excellence. The project provided that the Public works office supervised all the works while the Programming office looked for funding and drafted public calls. Over time, the Public works office changed several managers while the Programming office mostly kept the same staff until 2007. That was the reason why we successfully managed EU and regional funds. You need cohesion and distribution of competences within the administration, otherwise, it is not possible. (Appendix, Interview, Group A)

> There was a very capable manager, who later moved to the Province of Sassari, and now has an agreement with the Municipality of Alghero. He is an expert in funding schemes and did a great job. He was part of a team, which also included the city manager, who worked hard and obtained a great number of funds. (Appendix, Interview, Group D)

Although the staff of the sectors involved did not consist of an appropriate number of employees, the Municipality of Alghero, thanks to the dedication of its technical and administrative staff, managed to cope with the lack of people:

> I did not have many employees, however, I had good technicians. And so, let's say that it was possible to carry out these projects in the best way [. . .]. At Public works, I had a dozen people, including technical and administrative staff, but we were a balanced group of professionals, and it worked fine. (Appendix, Interview, Group B)
>
> Of course, the administration tried to get the most out of it, and as technical support, we tried to obtain funds whenever possible. [. . .] The regional government published calls that we had to stick to. At the Programming office, [. . .] we drafted projects to submit under those calls. The biggest problem was the limited time available because after being informed of the call, you had to draft the project and finalize it by obtaining approvals from the competent bodies. (Appendix, Interview, Group D)
>
> Despite having very few staff members [. . .], we managed to obtain most of the funds we applied to. The Region of Sardinia lost a lot of funds due to the lack of projects because, unfortunately, several administrations did not care about developing projects and submitting them for funding. (Appendix, Interview, Group A)
>
> Being new to EU funding, we had a tough experience because the process required a series of regular paperwork that took time and effort. Also, the administrative staff of the Municipality was perhaps not properly trained to complete the tasks [. . .]. (Appendix, Interview, Group C)

During those periods of funding, the political and the administrative players worked in synergy to make the most out of all funding opportunities:

> My goal was to seize any EU funding opportunity. That was the challenge. We also had planned several forms of co-financing, proportional to the opportunities available from EU programmes. We built a co-working environment because I created an internal Project design office. Officers and executives officers were not passive players in those projects: they were proactive. (Appendix, Interview, Group B)
>
> In ten years, Alghero received from the EU €40 to €50 million to regenerate the city. The Municipality was subject to a considerable stress. On the one side, it was a political decision, that is, it was connected to both the ability to launch a Project design office that would monitor the opportunities provided by the EU, and to the Public works' ability to carry out those projects, implement them and bring them to completion. The interaction between the political and the administrative bodies was continuous and depended on the ability of those [city] managers to motivate their staff, to replace employees who were not suitable for the job with employees who were. There was a daily routine to follow up with the offices. From an administrative point of view, a phone call by a city counsellor or the mayor to the *Soprintendenza* was more effective than the call of the surveyor or the engineer who monitored the project. (Appendix, Interview, Group C)

The Municipality of Alghero found it challenging to manage the relationships with the Region resulting from administrative uncertainty in managing funds. At first, such an uncertain situation was also a direct consequence of the inexperience of the municipal administrative apparatus, but later on, especially thanks to the collaboration between the administrations of the two institutions and to the establishment of an effective communication channel, it was solved by establishing specific rules:

> The relationship with the regional government was challenging because we were not very well prepared, being the first time for the Municipality to manage EU funds. We did not know all the EU directives properly and the ways funds had to be managed, we constantly had to seek help from regional coordinators. The regional government was quite helpful on this, so, in the end, we could still complete the process in the best way possible. [...] Let's say that the greatest difficulty was to prepare all the documentation for the final financial report and the audit. We had to transfer data to the regional government every two months, but in the end, everything went fine. [...] Let's say that the regional government's availability was good. (Appendix, Interview, Group A)

Degree of Structuring and Impermeability of Gatekeeping

The third dimension in the institutionalisation of the governance system relates to the (c) degree of structuring and impermeability of gatekeeping, particularly concerning the actors who are able to control, at least at the major nodal points, the political, legal and economic–financial resources of the decision-making process.

In the case of Alghero, the economic–financial management of funds was carried out by the Region and the Municipality. The Region was responsible for the management of EU funds, as well as for the actual transfer of funds, but it also played a variety of auditing functions by entering into specific agreements which secured comprehensive monitoring of the resources managed by the Municipality. Access to and the control of resources was, therefore, a privilege of those two players only, thus causing a high level of structuring and gatekeeping impermeability:

> The Region was responsible for the funding and deadline monitoring. (Appendix, Interview, Group A)
>
> Whenever we were assigned funds, the Municipality had to sign an agreement providing a whole set of constraints; failure to comply with them would result in funding loss. The main problem was probably meeting deadlines because unfortunately, the time frame was often very tight to implement the measures; time frames were hardly compatible with works, such as the restoration and recovery of historical buildings, which often involved delays and extensions. (Appendix, Interview, Group B)

Some aspects that allow assessing gatekeeping permeability relate to the control of technical–administrative and organisational resources of policy-making during the implementation phase. The policy has been characterised by a steady decrease in the level of gatekeeping permeability. During the first phase, the level was high, because initially, the municipal administration had listened to the needs of local commercial operators. After the first phase, through negotiation, the Municipality tried to meet the requests of the business owners, and then the municipal administration neglected the pressure exerted by the economic operators working in the area affected by the works:

> When we needed to start the works on the *Barcelona* promenade, also known as *Passeggiata Busquets*, we negotiated to remove all the kiosks from the promenade, and that was truly

> complex. To avoid damaging those businesses during summertime, the most profitable time of the year, we decided that the works should be carried out in winter. (Appendix, Interview, Group A)

Then, probably because of a lack of political resources, the level of gatekeeping impermeability declined rapidly. On various occasions, the dominant coalition of the policy network had given in to the pressure from different interest groups, such as environmental groups and business owners. There are two main reasons for this failure:

(a) The desire of the Municipality of Alghero to have a project approved that included constraints and changes for local businesses, who have not appreciated the choices made in the beginning. Due to 'political pragmatism', the Municipality, therefore, decided to accept the requests by redesigning the policy in some substantial parts.
(b) The scarcity of political resources from the Municipality of Alghero, who welcomed the requests due to the pressure exerted by movements which were supporting the dominant coalition:

> The *Busquets* project was launched by the left-wing ruling party between 1994 and 1998. But the project was approved by the Baldino City council. Later on, the project changed due to the lobbying of shopkeepers [. . .] and the kiosks were moved to the promenade, thus making a very large peninsula. (Appendix, Interview, Group C)
>
> The kiosks had to stay in the central *Rambla*, not on the promenade. They were quite happy to be granted a space next to the harbour. People had to walk on the bike lane because the kiosk keepers kept their stands on the pedestrian lane. (Appendix, Interview, Group E)
>
> The final part of the *Passeggiata Busquets* changed after the new City council was elected. [. . .] The project was amended by including the requests made by the environmentalists, and the lookout included in the project would then be built on the beach. Originally, the project included a star-shaped platform, nothing more than a terrace on the beach. [. . .] The star was rejected by political environmentalists within the dominant coalition; already during the election campaign, that work had been radically changed. (Appendix, Interview, Group E)
>
> There had been pressure from external parties, and the City council had changed the project based on external forces. The council had a close connection with the movements. Those movements influenced the political choice because they believed that the lookout to be built on the beach would result in less space for the people. So, they pressed the City council to withdraw the proposal. [They were] political movements; there was a local movement in Alghero that had supported [the mayor] during the political campaign. It was a strong movement that had somehow affected the implementation of the project. And the pressure of the environmentalists was very [strong] too. (Appendix, Interview, Group E)
>
> Regarding the [star-shaped veranda], we in the opposition claimed that the main resource was the sandy shore, the beach, and we continue to say so. We did the same for the port expansion. The greatest of Alghero's assets is the beach, especially because it is an urban beach. We are so lucky to have a stunning beach in our town, so any work preventing swimmers from using it for other purposes would not be [welcomed]. [. . .] It was an almost exclusively political move; in the end, there were more citizens against the star-shaped rather than in favour. (Appendix, Interview, Group C)

3.4.2.2 The Quality of the Urban Regeneration Policy

Degree of Responsiveness

By analysing the quality of the urban regeneration policy in terms of the (a) degree of responsiveness, the interviewees describe that originally the Old Town of Alghero was in a compromised situation, marked by a sharp deterioration and a major demand for regeneration:

> Before those maintenance works, the [non-degraded] area was limited only to four roads: *via Carlo Alberto*, *via Roma*, *piazza Civica*, and *via Gilbert Ferret*. The other parts of the Old Town were abandoned because the old hospital was crumbling. Those who lived in its surroundings suffered from highly degraded conditions. (Appendix, Interview, Group B)
>
> There were squares in the Old Town paved with tarmac, areas without pavement, truly degraded areas, which were not used either by tourists or by the local citizens because they were dirty and infested with weeds. (Appendix, Interview, Group D)
>
> The old hospital was abandoned. The *Quarter* was frequented by families in socially difficult conditions. It was isolated. The police station was in ruins and abandoned. [...] What was the Old Town like? It was a series of abandoned buildings and blocks. (Appendix, Interview, Group B)

Considering the existing degraded situation, the Municipality of Alghero wanted to prevent residents from leaving the area and fostering it as a commercial area:

> The Old Town is habitable; the only problem is to ensure that residents stop relocating. (Appendix, Interview, Group B)
>
> Public works lead to regeneration, but they also improve the quality of people's life. When you generate income when a place that used to be abandoned becomes a place of opportunity, because it is pleasant and people like to spend their time there, that becomes an economic asset, and I think that it is the best achievement for a policy-maker. (Appendix, Interview, Group A)

The implementation of the policy prompted a revitalisation process and the 'rediscovery' of a forsaken part of the city and to steady growth of the housing market. However, it generated clashing results: it led to a revitalisation of the economy of the area; at the same time, however, it determined the abandonment of the Old Town by the resident population and the increase of 'second houses':

> Meanwhile, along with the overall redevelopment of the Old Town, we have revitalized it. Perhaps even too much, I dare say, because operators have opened new shops only within the Old Town. This created a clash between the shopkeepers and the residents, [...] a structural and a cultural redevelopment because people move to the Old Town again. In addition, an economic redevelopment followed, because the revitalization of the Old Town fostered the local economy; this improvement was beneficial for all the citizens because the value of the existing real estate greatly increased. (Appendix, Interview, Group A)
>
> That is what happened. Following the restoration of the ramparts and the promenade, the Old Town was revitalized. I always remember that, until the mid-Nineties, the ramparts were a hiding spot for criminals because they were dark, no one could go there in the evening. It has become a place where people walk, eat while watching the sea, have fun, take their children for a walk. So, this process persuaded people to buy flats in the Old Town. A few years later, also the value of the flats in the city centre increased very much; so, it was a virtuous cycle. (Appendix, Interview, Group B)

Degree of Cooperation

As far as the second dimension of the quality of the urban regeneration policy is concerned, that is, the (b) degree of cooperation, since the very beginning, the policy was characterised by a bent for cooperation by the stakeholders interacting within the policy network. Every stage of the decision-making process was characterised by the use of collaborative methods and consultation, as well as an exact definition of the competences of each player involved:

> I must say that [. . .] the regional government has always been willing to provide support. We talked very much, and I spent a lot of time travelling to Cagliari. [Both] Mayor Marco Tedde and I had a great interest by regional managers and the officers of the Public works office. And so, we took advantage of all available opportunities. (Appendix, Interview, Group B)
>
> The Region did its part and, when necessary, we discussed the interpretation of clauses; the Region did not provide much training, but its staff teamed up to solve problems from time to time. (Appendix, Interview, Group D)
>
> [Relations were] close: in the end, discussions took place between people. When no strict rules had to be complied with, (e.g. regarding budget, deadlines), personal discussions made it possible to analyze issues, adjust measures, exchange ideas. So, there were phone calls and informal meetings. (Appendix, Interview, Group A)

The use of this method for the resolution of disputes also determined the absence of any conflict among the players that interact in the decision-making network. Although the players pursued divergent interests, the dominant coalition of the policy network succeeded in reaching an agreement in all circumstances, which would bring different interests together:

> Therefore, we signed MOU's to carry out specific activities because available funds were allocated mostly for specific purposes. (Appendix, Interview, Group E)
>
> Although at first, kiosk keepers did not want to collaborate, refusing to remove their kiosks, we found an agreement. We met kiosk keepers individually to reach an agreement. At the end, that did the trick. (Appendix, Interview, Group B)

Degree of Functional Congruence

As far as the third dimension of the quality of the urban regeneration policy is concerned, that is, the (c) degree of functional congruence, the case of Alghero, unlike that of Sassari, showed that the urban regeneration policy of the Old Town was highly interesting for private stakeholders, particularly for bar and restaurant owners, and kiosk keepers, who saw the opportunity to undertake business ventures in a regenerated Old Town:

> In principle, the objective was to regenerate and recover an abandoned piece of urban, architectural and historical heritage. But the result was much better because the recovery of history and culture also resulted in an unexpected economic development. All the activities that developed there and the flows of people persuaded several young entrepreneurs in Alghero to launch new businesses. (Appendix, Interview, Group C)
>
> The first area affected by the redevelopment was the perimeter of the Old Town, where new restaurants and bars started business. (Appendix, Interview, Group E)

The Alghero airport, in those years, became one of the destinations of Ryanair low-cost airline. This fostered tourism and resulted in an increased demand for real estate, mainly in the Old Town, creating a sort of 'real estate bubble':

> The regeneration policy revived the Old Town. Later on, new businesses were launched on the ramparts, especially restaurants, ice cream parlours, etc., as well as the promenade for [tourists]. So, without that redevelopment, no benefit would have come for local businesses. When low-cost travels became popular, also the demand for flats in the Old Town increased. It was a cyclical phenomenon, let's say it was a phenomenon tied to the economic cycle. (Appendix, Interview, Group A)
>
> With the arrival of Ryanair in Alghero, Europe got closer. Many Italian and foreign tourists enjoyed and appreciated the Old Town and began investing in real estate. (Appendix, Interview, Group C)

The interest of commercial operators, together with the major housing demand generated during that period, significantly sped up the whole policy process, which ultimately involved all private stakeholders, both business operators and citizens. This evolution also led to a change in people's habits, who began to spend time in the Old Town that previously had a marginal attendance:

> It was beneficial for the whole area: the population of Alghero and the tourists started to spend more time in the Old Town that had been regenerated. The result was outstanding from that point of view. In the summertime, there are so many people on the streets that taking a walk is almost impossible in the Old Town. And this is the case also in winter. (Appendix, Interview, Group E)

The start of that trend also generated negative aspects since the proliferation of shops and the massive attendance of tourists resulted in a worse quality of life, mainly at night and especially for the resident population. Therefore, residents started to leave the Old Town, which ended up being populated mainly by non-residents, in the summertime:

> Meanwhile, following the overall redevelopment of the Old Town, we managed to give a new life to it. Perhaps even too much, I dare say, because many businesses were shops were launched only in the Old Town. This affected the coexistence of businesses and residents. (Appendix, Interview, Group A)
>
> Once you promote a regeneration policy, you fix the road surface, the cobblestones, the lighting, street furniture, you increase the availability and usability of spaces. On the other hand, from an 'anthropological' point of view, the few residents who were still there decided to leave the Old Town because businesses expanded more and more; also, the noise increased, and the streets became overcrowded. Today, the Old Town is mainly inhabited by non-residents, by people who come here in the summertime; to be honest, there are very few residents. (Appendix, Interview, Group E)

3.5 Overall Evaluations of the Typical Case Studies

3.5.1 Evaluation of the Institutionalisation of the Governance Systems

As illustrated in the table below (Table 3.2), with regard to the first analytical dimension of the independent variable—that is, the (a) functional specialisation and interdependence maps among the actors of the dominant coalition— the case of Sassari can be considered as a *medium* degree. This is because in the first phase of the policy, the City Council followed a relational channel that we defined as 'privileged' and direct with the Ministry of the Environment, which excluded (even if temporarily) the regional government of Sardinia from the policy-making process. In fact, the principle of collaboration and transparency underpinned their relations with the *Soprintendenzas*. Later on, the competences for managing the ROP funds were fully transferred to a regional executive committee. In the second phase, there was an increase in the degree of functional specialisation and of interdependence maps that is considered to be a medium/high degree. The overall medium level is, therefore, the result of the average between the first and the second phases of the policy.

On the other hand, in the case of Alghero (Table 3.3), the empirical analysis led us to assess it as an overall *medium/high* degree. This is because the dominant coalition of the policy network, once gathered, clearly defined its degree of specialisation and

Table 3.2 Overall evaluation of the level of institutionalisation of the governance system in Sassari

Analytical dimensions of the institutionalisation of the governance system	Overall evaluation in Sassari
(a) Degree of functional specialisation and interdependence maps among the actors of the dominant coalition	Medium/low
(b) Degree of congruence of cognitive maps within the dominant coalition	Medium
(c) Degree of structuring and impermeability of gatekeeping	Medium/low
Level of institutionalisation of the governance system	
Medium	

Table 3.3 Overall evaluation of the level of institutionalisation of the governance system in Alghero

Analytical dimensions of the institutionalisation of the governance system	Overall evaluation in Alghero
(a) Degree of functional specialisation and interdependence maps among the actors of the dominant coalition	Medium/high
(b) Degree of congruence of cognitive maps within the dominant coalition	Medium/high
(c) Degree of structuring and impermeability of gatekeeping	Medium/low
Level of institutionalisation of the governance system	
Medium	

described its interdependence maps, considering that both the roles and communication channels were clearly defined.

Moving to the overall evaluation of the degree of the second analytical dimension of the independent variable, namely, the (b) congruence of cognitive maps within the dominant coalition, the case of Sassari can be considered as a *medium-level* degree (Table 3.2). Our study showed that even if there was a high congruence between the cognitive maps of political and administrative bodies within the municipal administration and between the Municipality and the *Soprintendenza*s, the same does not seem to apply to the cognitive maps of the Region and the Municipality.

Unlike Sassari, Alghero shows a *medium/high* degree of congruence of cognitive maps (Table 3.3). The reason for this evaluation is that we found a high degree of congruence between the cognitive maps of political and administrative bodies and a progressive improvement of relations with the Region of Sardinia.

Turning to the third analytical dimension—that is, the (c) structuring and impermeability of gatekeeping— even though the dominant coalitions, in both cases, at first showed a tendency to impermeability to the actions of interest groups, a *medium/low* degree can be assessed (Tables 3.2 and 3.3). This evaluation stems from the fact that the dominant coalitions of the policy network were unable to fully control the most important issues in the decision-making process, due to a lack of political resources, without being influenced by private players, particularly those who had a technical–administrative and organisational nature.

All this considered, the overall level of institutionalisation of the governance system in Sassari, as shown in the table above (Table 3.2), can be assessed as *medium*. The following table (Table 3.3) shows the overall level of institutionalisation of the governance system in Alghero, which can also be assessed as *medium*.

3.5.2 *Evaluation of the Quality of Urban Regeneration Policies*

As far as concerns the first analytical dimension of the dependent variable—that is, the (a) degree of responsiveness—as shown in the table below (Table 3.4), the case of Sassari attests to the presence of a *low* degree of responsiveness to the citizens'

Table 3.4 Overall evaluation of the level of quality of urban regeneration policy in Sassari

Analytical dimensions of the quality of urban regeneration policy	Overall evaluation in Sassari
(a) Degree of responsiveness	Low
(b) Degree of cooperation	Medium/low
(c) Degree of functional congruence	Medium/low
Level of quality of urban regeneration policy	
Medium/low	

Table 3.5 Overall evaluation of the level of quality of urban regeneration policy in Alghero

Analytical dimensions of the quality of urban regeneration policy	Overall evaluation in Alghero
(a) Degree of responsiveness	Medium/high
(b) Degree of cooperation	Medium/high
(c) Degree of functional congruence	Medium/high
Level of quality of urban regeneration policy	
Medium/high	

requests, due to the inadequacy of the results achieved for the regeneration of the Old Town.

On the other hand, the case study of Alghero shows a different situation. The policy designed and implemented by the Municipality of Alghero led to the regeneration of the area and, consequently, to the revitalisation of its economy. This was particularly the case for the real estate market that experienced an exponential increase. This increase was also followed by an increase in the number of 'second houses' in the area. Otherwise, that trend was not matched by an increase in the resident population, due to a worsening of living conditions determined by the presence of several businesses as a result of the urban regeneration policy. Therefore, the degree of responsiveness in the case of Alghero can be assessed as *medium/high* (Table 3.5).

Moving on to the evaluation of the second analytical dimension of the dependent variable—that is, the (b) degree of cooperation—both in the cases of Sassari and Alghero, the interactions among the institutional players, particularly among the regional government, the Municipalities and the *Soprintendenzas*, show a high degree of cooperation. The evaluation changes when we consider the relationship between public and private stakeholder. As evidenced by the empirical data, indeed, opposite trends are reported. As far the case of Sassari is concerned, the involvement of the private players by the Municipality was minimal and highly confrontational; so, in the case of Sassari, as shown in the table below (Table 3.4), the overall degree of cooperation among the players of the policy network can be rated as *medium–low*.

In the case of Alghero, instead, there was a high degree of involvement and collaboration of the private operators, both shop owners and private citizens, and as a consequence, the overall degree of cooperation between the players of the policy network can be considered as *medium–high* (Table 3.5).

Turning to the overall evaluation of the third and last analytical dimension of the dependent variable—namely, the (c) degree of functional congruence—the case of Sassari shows that the successful redevelopment of urban furniture and underground systems was not followed by an overall regeneration of the Old Town, especially in terms of housing and business. Except for very few areas, the Old Town of Sassari still remains highly degraded and poorly attractive for both economic investments and residential relocation. Therefore, the degree of functional congruence can be assessed as *medium–low* (Table 3.4).

On the other hand, the Alghero case shows that the urban regeneration policy of the Old Town was highly interesting for private stakeholders who invested in the improvement of the real estates within this area, together with other exogenous factors (above all, the fact that the Alghero airport became a national/international hub for low-cost airlines, like Ryanair), a fruitful real estate market and the revitalisation of commercial activities in the Old Town. However, those trends paradoxically lowered the number of inhabitants/residents within the Old Towns, as a large number of tourists and the high concentration of commercial activities reduced the quality of life of the residents, encouraging their relocation. Due to this whole set of dynamics, in Alghero, the degree of functional congruence can be considered as *medium–high* (Table 3.5).

Based on all the above considerations, the overall level of quality of the urban regeneration policy in Sassari can be assessed as *medium/low* (Table 3.4).

The above table (Table 3.5) shows instead the overall level of quality of the urban regeneration policy in Alghero, which can be assessed as *medium/high.*

References

Allulli M (2011) Pratiche partecipative e istituzionalizzazione. Tra ritualità e decision-making. Riv Ital Politiche Pubbliche 6(3):443–476

Allulli M, Tortorella W (2013) Cities in search of policy. The urban issue in the Italian national political agenda. Métropoles 12, https://doi.org/10.4000/metropoles.4654

Andersen HT (2001) The new urban politics of Europe. The area-based-approach to regeneration policy. In: Andersen HT, Van Kempen R (eds) Governing European cities. Asghate, Hampshire, pp 233–253

Andrew C, Goldsmith M (1998) From local government to local governance—and beyond? Int Polit Sci Rev 19(2):101–117

Atkinson R, Carmichael L (2007) Neighbourhood as a new focus for action in the urban policies. In: Smith I, Lepine E, Taylor M (eds) Disadvantaged by where you live? Neighbourhood governance in contemporary urban policy. The Policy Press, Bristol, pp 43–63

Bagnasco A (1990) La città dopo Ford: il caso di Torino. Bollati Boringhieri, Torino

Bighi S (2016) Le aree dismesse nella riqualificazione e nella rigenerazione urbana a Torino (1990-2015). In: Armano E, Dondona CA, Ferlaino F (eds) Postfordismo e trasformazione urbana. Casi di recupero dei vuoti industriali e indicazioni per le politiche nel territorio torinese. Istituto di ricerche economico-sociali del Piemonte, Torino, pp 107–140

Chiodini L, Milano R (eds) (2010) Le città ai margini: povertà estreme e governo delle aree urbane. Ministero del lavoro e delle politiche sociali, Roma

Christensen K (1999) Cities and complexity: making intergovernmental decisions. Sage, London

Contardi L, Moscato M, Ricci M (eds) (1999) Programmi di riqualificazione urbana, azioni di programmazione integrata nelle città italiane. Istituto Nazionale di Urbanistica, INU Edizioni

Couch C, Fraser C, Percy S (eds) (2008) Urban regeneration in Europe. Blackwell, London

Cremaschi M (ed) (2002) Politiche territoriali e programmi europei. Arch Studi Urban Reg, 75(3)

Del Genio V (2007) Le politiche di riqualificazione urbana e la fiscalità di vantaggio. Econ Ser 2 (2):341–360

Dente B (1990) Metropoli per progetti? In: Dente B, Bobbio L, Fareri P, Morisi M (eds) Metropoli per progetti. Attori e processi di trasformazione urbana a Firenze, Torino e Milano. Il Mulino, Bologna, pp 221–244

DiGaetano A, Strom E (2003) Comparative urban governance. An integrated approach. Urban Aff Rev 38(3):356–395

European Commission (1997) Towards an urban agenda in the European Union, https://eur-lex.europa.eu/legal-content/EN/TXT/PDF/?uri=CELEX:51997DC0197&from=it

Eurostat (2020) Eurostat data explorer, https://appsso.eurostat.ec.europa.eu/nui/show.do?dataset=urb_lpop1&lang=en

Evans G, Shaw P (2004) The contribution of culture to regeneration in the UK: a review of evidence. London Metropolitan University, London

Galdini R (2008) Reinventare la città. Strategie di rigenerazione urbana in Italia e Germania. Angeli, Milano

Governa F, Janin Rivolin U, Santangelo M (2009) La costruzione del territorio europeo. Sviluppo, coesione, governance. Carocci, Roma

Guala C (2007) Mega eventi. Modelli e storie di rigenerazione urbana. Carocci, Roma

Gualini E (2004) Regionalization as 'experimental regionalism': the rescaling of territorial policy-making in Germany. Int J Urban Reg Res 28(2):329–353

Gualini E (2006) The rescaling of governance in Europe: new spatial and institutional rationales. Eur Plan Stud 14(7):881–904

Gualini E (2017) Institutional capacity building as an issue of collective action and institutionalisation. Some theoretical remarks. In: Cars G, Healey P, Madanipour A, De Magalhaes C (eds) Urban governance, institutional capacity and social Milieux. Routledge, London, pp 29–44

Guiati F (2008) Rigenerazione urbana e accompagnamento sociale. Il caso di Via Artom a Torino. Celid, Torino

Indovina F (2012) Governare la città con l'urbanistica. Guida agli strumenti di pianificazione urbana e del territorio. Maggioli, Rimini

ISTAT (2019) Statistiche demografiche, http://demo.istat.it/pop2019/index.html

James P (2015) Urban sustainability in theory and practice. Routledge, London/New York

John P (2001) Local governance in Western Europe. Sage, London

Karrer F, Moscato M, Ricci M, Segnalini O (1998) Il rinnovo urbano. Programmi integrati di riqualificazione e di recupero urbano: valutazioni e prospettive. Carocci, Roma

Kaufmann D, Sidney M (2020) Toward an urban policy analysis: incorporating participation, multilevel governance, and seeing like a city. Polit Sci Polit 53(1):1–5

Leat D, Setzler K (2002) Towards holistic governance: the new reform agenda. Palgrave, Basingstoke

Lees L (2008) Gentrification and social mixing: towards an inclusive urban renaissance? Urban Stud 45(12):2449–2470

Maciocco G (ed) (1991) La pianificazione ambientale del paesaggio. Angeli, Milano

Maciocco G (2007a) Architecture environment and beyond. Skira editore, Milano

Maciocco G (2007b) Fundamental trends in city development. Springer, Berlin/Heidelberg

Maciocco G (2008) Urban landscape perspectives. Springer, Berlin/Heidelberg

Maciocco G, Camarda I (1991) Le dimensioni ambientali della pianificazione urbana. Angeli, Milano

Maciocco G, Clemente F (1996) La città in ombra: pianificazione urbana e interdisciplinarità. Angeli, Milano

Maciocco G, Pittaluga P (2001) La città latente. Il progetto ambientale in aree di bordo. Angeli, Milano

Mela A (2006) Sociologia delle città. Carocci, Roma

Moulaert F, Mehmood A (2020) Towards a social innovation (SI) based epistemology in local development analysis: lessons from twenty years of EU research. Eur Plan Stud 28(3):434–453

Nel·lo O (2010) The challenges of urban renewal. Ten lessons from the Catalan experience. Anál Soc 45(197):685–715

Novembre C (2011) Trasformazioni urbane e sviluppo locale. Elementi di competitività tra città. PhD thesis, Università degli Studi di Catania

OECD Territorial Review (2006) Competitive cities in the global economy, https://pure.mpg.de/rest/items/item_1233685_3/component/file_1233684/content

Padovani L (2000) Il recupero urbano. In: Indovina F, Fregolent L, Savino M (eds) 1950–2000. L'Italia è cambiata. Angeli, Milano, pp 352–355
Palermo PC (2002a) Apprendere da Urban: effetti immediati e incerti sviluppi di un'esperienza innovativa. Urbanistica 119:36–38
Palermo PC (ed) (2002b) Il programma Urban e l'innovazione delle politiche urbane, vol I. Angeli, Milano
Palermo PC, Savoldi P (eds) (2002) Il programma Urban e l'innovazione delle politiche urbane. Esperienze locali: contesti, programmi, azioni, vol II. Angeli, Milano
Pasqui G (2011) Un ciclo politico al tramonto: perché l'innovazione delle politiche urbane in Italia non ha funzionato. Territorio 57:147–156
Pierre J (2005) Comparative urban governance. Uncovering complex causalities. Urban Aff Rev 40 (4):446–462
Saccomani S (2004) Programmi complessi: una rilettura delle esperienze. In: Piemonte R (ed) Valutare i programmi complessi, Stamperia Artistica di Savigliano, Torino, pp 15–38
Saccomani S (2016) Urban regeneration and crisis. EURA conference: cities beyond notwithstanding the crisis. In: Torino
Sanna A (2011) Il piano strategico provinciale e gli strumenti di pianificazione del territorio. In: Maciocco G, Balestrieri M, Congiu T (eds) Il progetto urbano del territorio: la pianificazione strategica della Provincia di Sassari. Angeli, Milano, pp 25–34
Savino M (2006) Rammendare. In: Indovina F, Fregolent L, Savino M (eds) Nuovo lessico urbano. Angeli, Milano, pp 100–106
Sbetti F, Rossi F, Talia M, Trillo C (eds) (2013) Il governo della città nella contemporaneità. La città come motore di sviluppo. INU Edizioni, Roma
Secchi B (2005) La città del ventesimo secolo. Laterza, Roma-Bari
Stoker G, Mossberger K (1994) Urban regime theory in comparative perspective. Eviron Plann C Gov Policy 12(2):195–212
Tebaldi M, Calaresu M (2009) Valutare la democrazia: introduzione all'analisi della qualità democratica. Aracne, Roma
Tebaldi M, Calaresu M (2015) 'Democra-city': bringing the city back into democratic theory for the 21st century? City Terr Archit 2(1):13
van den Berg L, Braun E, van der Meer J (eds) (2018) National urban policies in the European Union. Routledge, London
Wolman H, Goldsmith M (1992) Urban politics and policy. Blackwell, Oxford

Chapter 4
Institutionalisation of the Governance System and the Quality of Environmental Protection Policy in Sardinia

Abstract This chapter is devoted to the empirical analysis of the urban regeneration policy for the typical case study of the *Area Marina Protetta della Penisola del Sinis – Isola di Mal di Ventre* (Sinis peninsula – Mal di Ventre island marine protected area) in the Mediterranean island of Sardinia, Italy (1993–2004). In the first section of the chapter, we present the environmental protection policy in Italy and Sardinia, addressing the relevant actors, objectives, resources and instruments embedded in the policy. In the second section, we test the degree of institutionalisation and the quality of the policy in the policy area within the selected case study. For this purpose, we highlight several analytical dimensions of the institutionalisation of the governance system as well as the quality of the territorial policies. In the third section, we perform an overall assessment of the case study, paying particular attention to evaluating the institutionalisation of the governance system and the quality of each observed territorial policy.

Keywords Environmental protection policy · Sinis peninsula – Mal di Ventre island MPA · Sardinia

4.1 Environmental Policies and Environmental Protection Policies

An environmental policy may be defined as '[. . .] that complex of interventions implemented in order to govern human activities that reduce the availability of natural resources or reduce their quality or capacity to be enjoyed' (Lewański 1997, 281). According to many scholars (see Lester 1995 for a comprehensive review on the topic), this type of policy can be better understood within the broader context of reformist public policies promoted in Western countries from the 1960s onwards, which have been based on a belief in the capacity of the political system and international organisations to bring about large-scale modifications in relation to social objectives (Haas et al. 1993).

© Springer Nature Switzerland AG 2021

M. Tebaldi, M. Calaresu, *The Quality of Territorial Policies in Europe's Periphery*, Urban and Landscape Perspectives 22, https://doi.org/

In the course of the last two decades (2000–2020), scientific evidence for the progressive deterioration of the planet's environmental conditions has caused great concern among various relevant academic communities, from climatologists and urban planners to political and social scientists (Dessler and Parson 2019), providing a wakeup call across ever-broader segments of (global) public opinion (Stoutenborough 2020), above all among younger generations (e.g. student movements involved in the *Fridays for Future* campaign). In addition and even though environmental policies are principally situated within an 'intergovernmental context [...] [which] often makes it difficult to identify the specific responsibilities for the performed activities' (Lewański 1997, 31), many sovereign states in Europe have invested considerable political, legal, communicational, administrative and financial resources in this policy area (Jordan 2005), also with reference to the peripheral regions of the EU (Baker et al. 1994; Yearley et al. 1994).

Within the context of environmental policy, the so-called environmental *protection* policies today are particularly important, to the extent that environmental conservation represents a major interest of many governments (see, for example, the Paris Climate Treaty in Bodansky 2016).

The consequence on the juridical plane is that environmental protection measures bring about more or less substantial publicist nature restrictions on '[...] economic activities and subjective rights' (Garzia 2011, 67). From this point of view, environmental protection policies can be considered a regulatory-type policy (Lewański 1990; Lowi 1964, 1972). Firstly, the mechanism on which they are based is in fact represented principally by positive or negative normative prescriptions promulgated by public authorities in relation to private entities, which are subject to orders and controls. Secondly, such policies are characterised by an impact that conditions the behaviour of individuals or specific categories of entities, enforcing respect for rules via sanctions and/or restrictions on freedom of action and modifying the range of permissible behaviours through the exercise of authority (Regonini 2001). Thirdly, environmental policies affect not just individuals or representatives of limited interests but also organised and opposing interest groups, which through conflictual or cooperative strategies bring about coalitions, which dominate entire sectors of the economy and society (Bulsei 2005, 18).

It is important to highlight that in addition to regulatory characteristics typical of environmental protection policies, there are also characteristics associated with distribution (through subventions or grants) and redistribution (reallocation of costs and benefits among particular social categories or territorial zones), as well as, in some cases, of constituent policy.

Lastly, it has become increasingly clear to political science scholars that regulatory measures are not a sufficient condition for obtaining the results that public policy repeatedly promises to realise. From the 1980s onwards, many authors have felt the need to understand more deeply the processes which lead to the formulation and implementation of environmental and environmental protection policies and to consider the effectiveness of these policies (Bulsei 1990, 2005; Dente et al. 1984;

Leroy and Crabb 2012), through a series of studies designed to measure '[...] the degree to which environmental [and environmental protection] policy objectives have been achieved' by verifying '[...] how much the outputs intended by the policy itself have actually been produced' (Lewański 1997, 68).

4.2 Environmental Protection Policies in Italy: Actors, Objectives, Resources and Tools

In Italy, the implementation of a *specific* environmental policy can be traced to the middle of the 1960s. Before then, only a few protection measures in favour of specific environmental assets had been put in place within the context of other public policies (Lewański 1997, 41). In the middle of the 1980s, with the establishment of the Ministry of the Environment (Law 349/1986), scholars noticed a simultaneous advance in terms of both quality and quantity: in a quantitative sense, there was an increase in output, both in terms of the regulatory sphere and in terms of the public funding dedicated to the environment; in a qualitative sense, a new government body had emerged which was capable of promoting its own interests in the environmental field and which had concentrated into one body formal responsibilities which had previously been dispersed.

Considering Italy's geographical location (situated in the middle of the Mediterranean Sea, between the Balkan and Hellenic peninsulas to the east and the Iberian Peninsula to the west, North Africa to the south and Continental Europe to the north, separated from the rest of Europe by the Alps—as well as its mountainous contours), Italy has 7900 km of coastline; two major islands, Sardinia and Sicily; and a great variety of landforms determined both by the nature of its landmass and the action of the sea—the policy subsystem associated with the sea and coastal protection has fundamental importance within the spectrum of environmental protection policies overall.

The loss of biodiversity and the alteration of natural habitats constitute one of the major environmental problems of our time, with a deterioration over the last decades of the century above all in relation to the marine and coastal environment (see, for example, Worm et al. 2006). There is a variety of reasons for this. Without purporting to be exhaustive, it is possible to mention the excessive exploitation of fish stocks, the increase in maritime traffic, pollution in its various forms and the more recent complex phenomenon of climate change (Haas 1989; Boehmer-Christiansen 1994a,b; Christie et al. 2005; Airoldi and Beck 2007; Owens 2010; Levin et al. 2012; Van Slobbe et al. 2013; Kannen 2014). To try to solve these kinds of problems, there are various types of instruments that it is difficult, if not impossible, to frame synthetically, both from a legal–normative point of view and from the point of view of policy instruments, in a single homogeneous category.

In terms of the present study, we can say that protected areas constitute one of the principal measures used by nation-states and international non-governmental organisations competent in particular in the field of marine biodiversity protection. The establishment of a marine reserve (Ferrajolo 1999; Abrami 2000; Weible and Sabatier 2005; Marino 2011; Fenberg et al. 2012; Edgar et al. 2014) is designed to find a balance between potentially contradictory values, for example, the economic exploitation of fish stocks and the enjoyment by tourists of the area in question, as well as the protection of the natural environment (Caveen et al. 2013, 2014). In this sense and apart from the purely regulatory aspects, the relationship between humans and their natural environment is at the centre of environmental protection policy (De Santo 2013; Leenhardt et al. 2013; Urquhart et al. 2013; Cárcamo et al. 2014; Potts et al. 2014; Katsanevakis et al. 2015). These areas can be then defined as '[...] those parts of the sea and coasts [...] in which the conservation of natural heritage and biodiversity is guaranteed through the imposition of a juridical regime differentiated from the normal juridical regime' (Garzia 2011, 10).

In Italy, the first legislative measure providing for integrated protection of marine areas was introduced with the adoption of Law 979/1982 'Measures for the Defence of the Sea', which dedicated an entire section, namely, Sect. V, to 'marine areas to which application of protection seems opportune' (Art. 26), including protection for the Sinis peninsula – Mal di Ventre island, in the Municipality of Cabras area of interest (Art. 31). The same law provides for the assignment of funds for the creation of these reserves (Art. 32) and was created with broader aims than those contemplated in Sect. V. In fact, its objective was to prevent large-scale marine pollution or at least limit its damaging effects on marine resources. The most innovative aspect of the law was the introduction, for the first time in Italy, even if only in programmatic terms, of a perspective attributing greater importance to the preservation of the natural environment than to its economic exploitation. MPAs incorporated a vision of the marine environment as an asset which needed to be protected by '[...] providing special protective regimes to those areas of particular interest from a natural, scientific, ecological, cultural or economic point of view' (Ferrajolo 1999, 451). This law, apart from defining the aims of the reserves (of a scientific, cultural or economic character), their regulation and their spatial extent, through Art. 27 also defined the protective regime for MPAs, specifying which activities would be prohibited or subject to restrictions or permits.

Following the promulgation of the above-mentioned national law, there was a period of inactivity in relation to this subject by the then responsible Ministry of the Merchant Navy, which was transformed in 1986 into the Ministry of the Environment. Moreover, it was, in fact, this Ministry which in practice began to apply the regulations which had been approved previously, while Law 349/1991 provided the first systematic framework governing the protection of nature by the State for both the terrestrial and marine environment. According to Art. 2 of this law, protected natural areas were classified into national parks (par. 1), regional parks (par. 2) and natural reserves (par. 3), while a category independent of the three preceding ones was inserted into par. 4 distinguishing protected areas as identified in the Geneva Protocol and those already defined in Italian Law 979/1982. In terms of the legal norms governing protected areas, Art. 19 prohibited all '[...] activities which may compromise the protection of the natural characteristics, which are the object of protection and the aims for which the area was established' (Ferrajolo 1999). The legislation also provided for the adoption of more detailed internal regulations specific to each reserve established, to be approved by a decree of the Minister of the Environment.

The MPA management framework provided under Art. 19 remained substantially unchanged with respect to the law establishing the Ministry of the Environment, which in turn had amended Law 979/1982. The implementation of this law was entrusted to the Central Inspectorate for the defence of the sea, which also relied on support from the port authorities. In some cases, there were provisions, through special agreements determined by the Minister of the Environment, for the law to be administered by public bodies, scientific institutions or recognised institutions (Ferrajolo 1999). Art. 36 of the law also identified further 26 areas[1] for the establishment of possible marine parks or marine reserves ('a soon to be established marine areas') in addition to those already provided by Law 979/1982. Additionally, the Council for the Defence of the Sea was given the power to select further possible areas within the national territory.

The next step took place with Law 426/1998, particularly in relation to MPAs. The principal aim was to address the limits in the application of the previous norms within the sector regarding developments within international norms.

One clear innovation within the financial context was that this law allowed for the provision of the necessary funds (in the 1998–2000 period) to establish those areas identified in the previous legislation. The law also introduced changes in the investigative phase, by abolishing the Council for the Defence of the Sea and entrusting this phase to a special technical secretariat established within the Ministry of the Environment and by transferring responsibility for the marine areas from the port authorities to public bodies, scientific institutions or environmental associations (a solution already contemplated in Italian Law 394/1992, albeit only as a possible eventuality). Apart from this, the surveillance roles which had at first been entrusted exclusively to the port authorities were now shared with the municipal and local police forces. Finally, the law sought to promote all forms of economically sustainable activity in the protected marine areas as well as the financing of all forms of initiative concerning environmental information, education and training.

After the 1998 Law, Law 83/2001 and Law 179/2002 were passed, leading to growing complexity in the relevant legislative framework. An opportunity for a comprehensive legislative reorganisation was provided by the passing of Law 308/2004, which contained among various items one relating to the administration of protected areas. The aim of this measure was to integrate and coordinate the administration of national, community and international regulations concerning the

[1]The MPAs identified by the law as 'available areas' (Art. 31 of Law 979/1982 and Art. 36 of Law 394/1991, as integrated by later dispositions of Law 344/1997, Law 426/1998, Law 388/2000 and Law 93/2001) were established on the basis of a special inquiry. Consequently, the establishment of an MPA entails, before anything else, the identification of an area of sea indicated as an 'available marine area' (Marino 2011).

Fig. 4.1 Underwater parks and MPAs in Italy.
Source: Adapted from the Ministry of the Environment (2019)

safeguarding of terrestrial and marine biodiversity. However, this delegation of authority did not produce any result, so that matters relating to protected areas were excluded from the 'Single Environmental Act' of 2006 (Garzia 2011, 15).

By the end of 2019 (the closing date of this analysis), 2 underwater parks and 27 MPAs had been established, taking in a total of approximately 228,000 hectares of sea and approximately 700 kilometres of coastline in Italy (Fig. 4.1), generally divided up in terms of typologies identified as zones of types A, B and C (Marino 2011).

Type A zones are defined as 'full reserve' zones, where there is an absolute ban on activities that could cause damage or disturbance to the marine environment. These represent the 'core' of the marine reserve. In these zones, which are usually small in area, the only activities usually allowed by the selected commissioner are those of service or scientific nature (e.g. access or swimming is not allowed).

Type B zones are defined as 'general reserve' zones. In these zones, a series of activities are permitted according to criteria determined by the selected commissioner, provided they have a limited environmental impact and allow for the sustainable enjoyment and use of the environment. These activities must be compatible with the resource, and the number of users and method of use is controlled (e.g. limits are placed on fishing and the types of equipment used, as well as the speed of transit within the area). Type B zones, in most cases, are large compared to the total area.

Zones of the so-called 'partial reserve' type C zones represent a buffer zone between the higher natural value areas and the sectors outside the marine protected area, where, in addition to activities permitted in other zones, other forms of sustainable enjoyment and use considered of modest environmental impact are also permitted. Generally, the activities allowed in these zones correspond with those permitted in types A and B zones, but with less restrictive conditions. In general, the greater part of the protected area falls within type C zones.

All three typologies are delimited by geographical coordinates and marked on maps attached to the decree establishing each protected area. Depending on the area being safeguarded, activities may be totally prohibited or subject to authorisation by the selected commissioner, issued after an appropriate evaluation of the activity requested and the prejudice which might be caused to the environment or particular habitat subject to conservation.

The regulations (set out in Law 979/1982, Art. 28; Law 394/1991, Art. 19, Subsection 6; and Law 93/2001, Art. 8, Subsection 8) represent the most important constraints since they constitute the normal form of regulations for permitted activities, prohibitions and waivers, according to the degree of protection necessary to safeguard ecosystems within a marine reserve. These regulations are proposed by the selected commissioner (subject to the reviews of the reserve commission) and then approved by the Ministry of the Environment. Thus, this Ministry has formal responsibility for controlling marine reserves activities since approval for the various activities allowed within different zones of the reserve rests with it (Garzia 2011).

The most crucial element in the system regulating activities within the marine reserves is Art. 19, Paragraph 3, of Law 394/1991, which sets out explicitly that 'within marine protected areas activities are prohibited which may comprise the safeguarding of the characteristics of the environment which are the object of protection and the reason for which the area was established' (e.g. any form of discharge of solid or liquid waste; the capture, collection or damaging of animal or vegetable species; as well as the removal of minerals or archaeological remains). It is also the responsibility of the administering bodies of the various marine protected areas, based on the nature and the individual socio-economic activities of the relevant territory, to determine further types of prohibition not provided by Art. 19 or, if appropriate, to allow for possible exemptions. The instrument used for this objective is the implementation and organisational regulations, which regulate in detail the activities permitted within the MPA.

To summarise, apart from the national legislation referred to, the powers provided for Italian MPAs are based on four distinct elements: (a) the ministerial decree by which they are established as well as any later modifying decrees, which determine the name and boundaries of the area; (b) the disciplinary regulations, which identify the protection zones and list the activities which may be carried out within them; (c) the implementation regulations, which define in detail how activities specified in the disciplinary regulations may be carried out; and (d) the organisational regulations, which define the organisational structure and in general the procedures under which the MPA is to function (Guidetti et al. 2008).

4.3 Environmental Protection Policies in Sardinia: Regional Planning Tools

Within the panorama of environmental protection, with particular reference to the safeguarding of the sea and coastline, the island of Sardinia is one of the regions of Italy with the highest level of protection, at least formally. Of the 48 marine areas identified in the national legislation referred to (Law 979/1982 and Law 394/1991), 10 are situated along the Sardinian coastline.

In Sardinia, the fundamental principles for the establishment and administration of national protected natural areas are regulated by Law 394/1991, while the principles for the establishment and administration of regional protected areas are contained in Regional Law 31/1989 ‘Norms for the establishment and administration of parks, reserves, and natural monuments, and areas of particular natural and environmental importance’ which ‘[. . .] define the general aims of conservation, recuperation and promotion of the biological, natural and environmental estate of the Sardinian territory’. This regional law sets out the necessary procedures for establishing a protected area and identifies the instruments for planning its sustainable administration. The protected area is entrusted to bodies with responsibilities within Sardinia or to consortiums of such bodies. The surveillance of the area is entrusted to the Forestry Corps and Environmental Guard. The parks, reserves, natural monuments and areas of natural significance and Sites of Community Importance (SIC) collectively make up the regional ecological network.

Law 979/1982 identifies areas for establishing marine reserves in Sardinia, Capo Caccia – Isola Piana, Penisola del Sinis – Isola Mal di Ventre, Golfo di Orosei – Capo di Monte Santo and Tavolara – Punta Coda Cavallo, while Law 394/1991 lists (either together with or separately from adjoining terrestrial protected areas) Arcipelago di La Maddalena, Capo Testa – Punta Falcone, Isola dell'Asinara, Isola di San Pietro, Capo Spartivento – Capo Teulada and Capo Carbonara.

The MPA, which is the focus of the present study, is the coastline in front of the Sinis–Montiferru Regional Park, which adjoins the Regional Reserves of Pauli Maiori, S'Ena Arrubia and the Marceddì, Corru S'Ittiri and San Giovanni wetlands. The idea of establishing a reserve first appeared on the council agenda of the Municipality of Cabras with Resolution 82/1988. Through this resolution, the council announced as one of its priorities obtaining direct administrative responsibility for the area ('selected commissioner' hereafter) from the national ministers, with the objective of creating investment in tourism and regulating fishing. Scientific studies necessary for understanding the environmental characteristics of the area and problems relating to the regulation of activities in the area were deferred until a subsequent phase.

In the same year, by Resolution 166/1988, a first proposal was approved by the City Council for defining boundaries and zoning of a marine natural reserve (to include only territory belonging to the Municipality of Cabras), which was to be submitted to the Minister of the Environment. The proposal as submitted was evaluated unfavourably by the Minister. Later communications between the ministers, the Sardinia Region (which in this preliminary phase remained on the sidelines even though it was kept informed via official communications of the procedures initiated by the local community), and the Municipality of Cabras took place without particular difficulties apart from those associated with technical corrections relating to the definition of boundaries and zones within the protected area. As a result of interactions with the Minister, the Municipality of Cabras in Resolution 194/1990 expressed itself favourable to the establishment of the Sinis peninsula – Mal di Ventre island marine natural area on two conditions: firstly, that a further revision of the boundaries and protection measures would take place after it was established and, secondly, that the organisational structure would identify the Municipality of Cabras as having sole direct control of the administrative functions of the reserve.

Despite the positive opinion of the municipality, the process for establishing the reserve was held up by a lack of response from the central authorities, who were engaged in a partial but significant reformulation of the regulatory framework referred to previously.[2] In this period, which we could describe as one of the transitions, the institutional players who were most interested in the Sinis peninsula – Mal di Ventre island marine reserve tried to put the establishment of this area at the centre of the political agenda through a series of municipal and parliamentary questions directed at the ministers responsible.

The procedural pathway was slowed down even further by the Sardinia Region—which when requested by the Ministry of the Environment to provide a formal opinion on the matter, as required under the law, instead requested that the region's

[2]During this phase, Framework Law 394/1991 and Law 37/1993 were promulgated, which modified in part the previous Law of 1982.

formal opinion be replaced by a national–regional agreement. The arm wrestle between institutional actors on this point was won in the end by the region. The stalemate only ended on 22 April 1997 (nearly 10 years after the first council resolution) when an agreement between the Ministry of the Environment and the Sardinia Region was signed, as a substitute for the formal opinion legally required from the region. On the basis of this agreement, the Municipality of Cabras met to vote Resolution 29/1997, which led to the signing of the Convention of 5 February 1998 between the Ministry of the Environment and the Municipality of Cabras, with the agreement of the Sardinia Region.

The municipality was able in this way to achieve its goal of sole administration of the area, including any economic and financial resources provided by the Ministry or by higher bodies (such as the EU) following the establishment of the reserve.

The administrative system brought into being through the creation of the MPA included the following roles: (a) the chair, a role entrusted to the mayor of the Municipality of Cabras, who acts in this role by virtue of his elected mandate; (b) the director, who acts as managing director and who was chosen and appointed by private agreement by the selected commissioner, subject to approval by the Ministry of the Environment; (c) the reserve commission, comprising a chair and seven members, established within the selected commissioner by decree of the Minister of the Environment as a consulting/controlling body supporting the managing director; (d) the secretariat, made up of technical personnel; (e) administrative personnel from within the Municipality of Cabras; and (f) any external staff appointed by special agreement.

In the years since the establishment of the reserve, its boundaries and zoning have undergone numerous changes. Leaving aside the configuration of the MPA pictured in the map below (Fig. 4.2),—representing the boundary and zoning of the MPA under analysis as set down in the 2011 Ministerial Decree, which is outside the time period covered in this study—it is possible to analyse the timeline of the changes in the different configurations of the Sinis peninsula – Mal di Ventre island MPA.

Over a period of 6 years, the following changes have taken place: the formulation of a new ministerial decree dated 6 September 1999 (in a coordinated text which completely replaces the preceding decrees of 12 December 1997 and 22 July 1999 regarding the establishment of the MPA) and the modification which took place with ministerial decree dated 17 July 2003, which defined new reserve boundaries and zones. Configuration (a) was proposed by the National Agency for New Technologies, Energy and Sustainable Economic Development (ENEA hereafter), which after researching an area of sea-facing the Sinis peninsula, produced in 1990 a document included in addition to the study results, as well as a proposal for new boundaries and zones.

The proposal included areas of natural and conservation interest but excluded areas that were mainly of economic interest. Configuration (b) shows the 1997 boundaries. This configuration assured the safeguarding of areas of conservation interest identified by the ENEA and included areas of economic and landscape

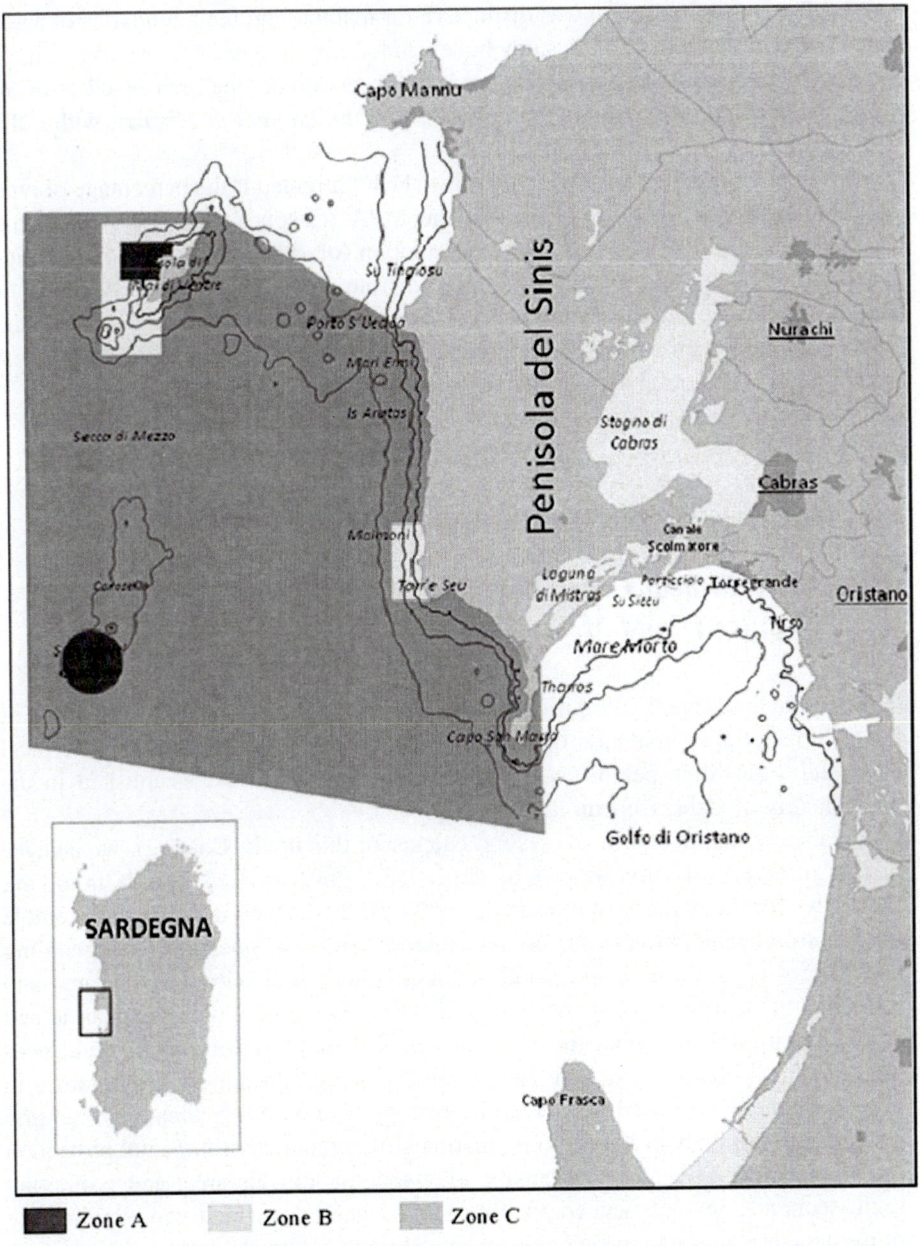

Fig. 4.2 Configuration of the Sinis peninsula – Mal di Ventre island MPA
Source: Adapted from Marra et al. (2017)

interest (bounded by type C zone) not included in the previous proposal. Consequently, in the Ministerial Decree of 1997, the configuration adopted covered a much larger area than initially proposed by the scientific agency.

Finally, configuration (c) was distinctive for maintaining the boundaries of areas already identified in the ENEA study but excluded from full protection two localities of particular importance for environmental conservation: 'the area of cliffs at *Su Tingiosu*, to the north, and the area overlooking the *Laguna di Mistras*, within the Gulf of Oristano' (Massaro 2010, 23).

The final important aspect which needs to be highlighted is the percentage of type A zone compared to the total area of the MPA (Casola et al. 2014). With the boundary changes between the first configuration (of 1997) and the second configuration (of 2003), type A zone was reduced from 3.6% to 2%. Type A zone full reserve areas represent the central or 'core' zone of the reserve and include '[...] the most important sector in terms of protection objectives, as they have the highest value in terms of conservation objectives' (Casola et al. 2008, 24).

4.4 Analysis of One Typical Case Study

4.4.1 The Typical Case Study of Sinis Peninsula – Mal di Ventre Island MPA (1993–2004)

This section describes the empirical analysis performed on the institutionalisation of the governance system of environmental protection policy and its quality in a typical case study: the Sinis peninsula – Mal di Ventre island MPA, established in the Municipality of Cabras, Sardinia, Italy (Fig. 4.3).

As already mentioned in the second chapter of this book (Chap. 2), we analyse marine reserve protection as part of the broader governmental agenda involving 'Measures for the defence of the sea' (Law 979/1982), which in particular promote the environmental protection of the marine area in question; safeguarding and valorising of the biological and geomorphological resources of the area and restocking of fish; increasing awareness of the ecology and biology of marine and coastal environments within the MPA and of the unique environmental and geomorphological characteristics of the area; introducing educational programmes to improve general knowledge in marine ecology and biology; creating scientific research programmes in the ecology, marine biology and environmental protection sectors, in order to achieve systematic knowledge about the area; and promoting socio-economic development compatible with the natural and landscape significance of the area, especially by privileging local traditional activities. In terms of promoting development compatible with the previously mentioned aims, the establishment of infrastructure for tourist transport would need to define regulations that specifically facilitate tourist transport preferably operated by residents within the territory. Our analysis covers, for this case, a period between 1993 and 2004.

As illustrated in the table below (Table 4.1), the study is conducted by examining the following: firstly, the independent variable, namely, the institutionalisation of the governance system, aiming to test the degree of its three dimensions

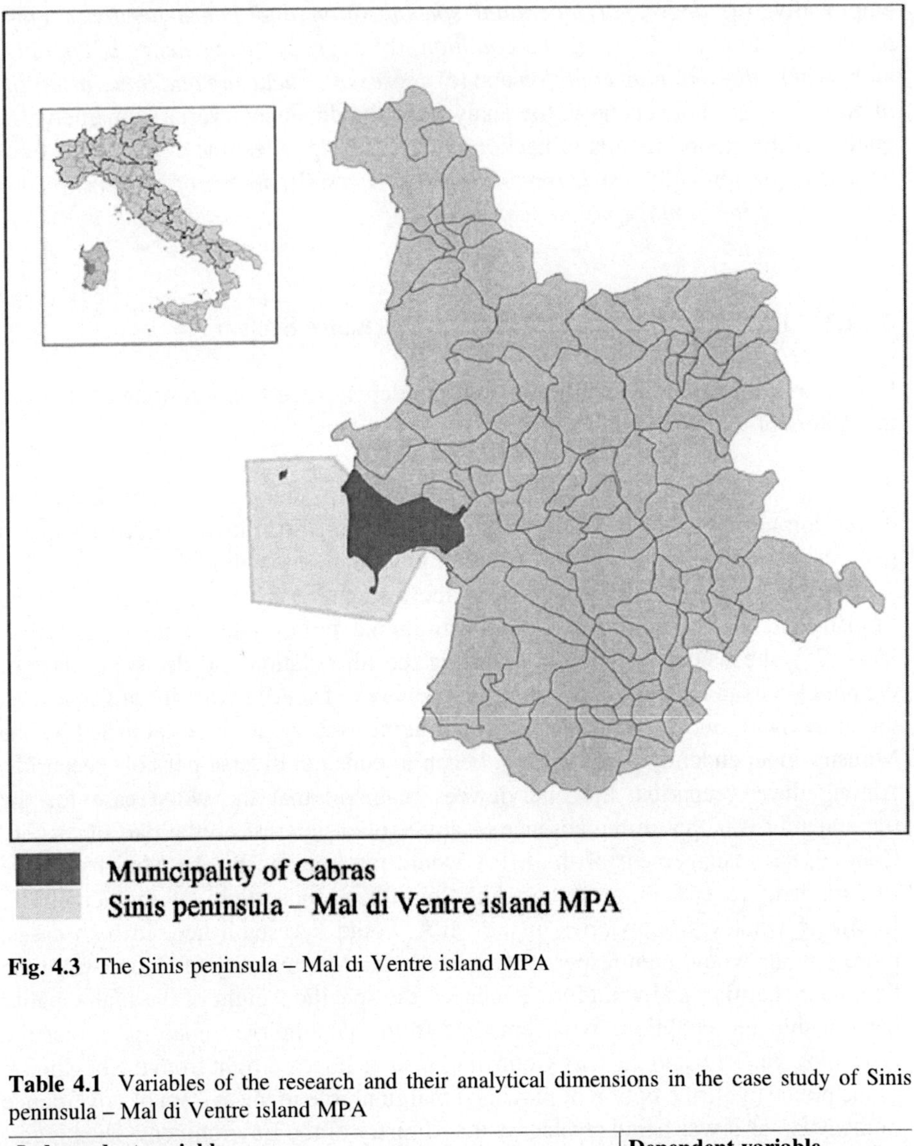

Fig. 4.3 The Sinis peninsula – Mal di Ventre island MPA

Table 4.1 Variables of the research and their analytical dimensions in the case study of Sinis peninsula – Mal di Ventre island MPA

Independent variable	Dependent variable
Institutionalisation of the governance system	*Quality of the environmental protection policy*
Analytical dimensions	**Analytical dimensions**
(a) Degree of functional specialisation and interdependence maps among the actors of the dominant coalition	(a) Degree of responsiveness
(b) Degree of congruence of cognitive maps within the dominant coalition	(b) Degree of cooperation
(c) Degree of structuring and impermeability of gatekeeping	(c) Degree of functional congruence

empirically, (a) *degree of functional specialisation and interdependence maps among the actors of the dominant coalition*, (b) *degree of congruence of cognitive maps within the dominant coalition* and (c) *degree of structuring and impermeability of gatekeeping*, and, secondly, the analysis of the dependent variable, namely, the quality of the environmental protection policy, aiming to test the degree of its three dimensions empirically, (a) *degree of responsiveness*, (b) *degree of cooperation* and (c) *degree of functional congruence*.

4.4.1.1 The Institutionalisation of the Governance System

Degree of Functional Specialisation and Interdependence Maps Among the Actors of the Dominant Coalition

As far as the institutionalisation of the governance system is concerned, the first dimension to be analysed relates to the (a) degree of functional specialisation and interdependence maps among the actors of the dominant coalition.

The empirical analysis conducted in the present case leads us to consider the Ministry of the Environment as the institutional policy entrepreneur (Lewański 1997, 95), the whose declared goal during the formulation and decision phase of the policy was to structure a strong policy network. The Municipality of Cabras, on the other hand, once it was clear that the marine area would be established by the Ministry independently of its willing, began to consider diverse possible scenarios. Among these scenarios, the interviewees maintain that the worst case for the municipality was that in the absence of any explicit interest on the part of its City Council, the management of the MPA would be assigned by the Ministry of the Environment to environmental organisations or a consortium of municipalities, following whatever boundaries of the MPA would be established. In both cases, those options would then respectively exclude the Municipality of Cabras from the dominant coalition and would have reduced the specific weight of the municipality itself within the coalition. What emerges from our empirical analysis is that the governing majority and the City Council of Cabras rejected right from the beginning of the policy cycle the option of playing a marginal role in the system of governance of the MPA as it was being created by the Ministry of the Environment. The biggest concern of the political class in Cabras at that time was that there would be a repetition locally of what had already happened with the Miramare Marine Reserve (the only example in Italy in the early 1990s of an area entirely entrusted by the government to the World Wildlife Fund (WWF) as an environmental association) and which ways weighed down by all of the environmental, landscape and natural restrictions potentially provided in the establishing law of the MPA:

> We thought that following the same path the Miramare marine reserve had followed with the WWF would not be appropriate [for us], because it had become an elite MPA. [...] In Cabras, we all agreed that it was necessary to make sure that, in particular, at the very beginning, the management [of the area] was entrusted to the Municipality [of Cabras]. And that was because we did not see any better solutions [.] Environmentalists are always too

> fundamentalist [.] So we thought that the Municipality of Cabras, especially at that time, was the best-qualified body to manage the [forthcoming] protected area. (Appendix, Interview, Group B)

Given the major interests at stake connected to the establishment of the MPA in terms of local development—above all in relation to the tourism and fishing sectors—the objective of the Municipality of Cabras was to refrain from the safeguarding restrictions being applied too strictly and without the involvement of the representative institutions of the local population. It was thought necessary for the municipality to act on its own and to anticipate developments in an attempt to obtain a direct and exclusive mandate from the Ministry of the Environment about the MPA:

> The fact that the request [to institute an MPA] was filed by us, as a Municipality, was welcomed with great satisfaction by the Ministry of the environment. Things were different compared to the MPA of Ustica, [where] the final agreement was signed [when someone] came from the Ministry in Rome by helicopter, and within half an hour, the deal was sealed in Sardinia [.] This means that the citizens found themselves [. . .] with a brand-new MPA with new restrictions without any discussion [.] In Cabras, as a Mayor, I did not want this to happen, [. . .] so that our town did not suffer a restriction of any kind without the citizens being aware of it. (Appendix, Interview, Group A)

During the period in which the policy was being established, the local administrators—independently of their various party affiliations—took on responsibility for direct dialogue with the Ministry of the Environment at a national level with a view to acquiring, through the delegation mechanisms provided in the establishing law, the most central role possible in terms of functional specialisations within the dominant coalition, even at the cost of losing their power of veto over the Ministry to decisions about the zoning of the MPA:

> In 1988, [. . .] the Cabras City council had shown a great deal of attention to the creation of the MPA [.] The establishment of a reserved area gained quite a lot of support within the council. I do not remember that there were any particular complaints about it [.] Regarding the zoning of the area, the City council expressed a favourable opinion [about the governmental proposal], as long as the management of the MPA would be entirely entrusted to the Municipality of Cabras. (Appendix, Interview, Group B)

Almost all of the city councillors who were interviewed state that the Municipality of Cabras had an indirect rather than direct interest in putting the environmental issue on their political agenda, which was a consequence of obligations under the national law referred to. Our empirical research shows that the desire to play a role in the management of the MPA was not the result of a bipartisan political agreement designed to initiate an environmental protection policy, but rather was the result of 'anxiety' on the part of the political class of the time about 'becoming the passive subjects of decisions by others' or even 'just having to make decisions in concert with other local authorities':

> [The interest in establishing the MPA] resulted from the adoption of national law. Before that, [. . .] the MPA was not one of my goals, [and] I don't think anyone else in Cabras had this in mind. To tell the truth, as [a former] Mayor of the city of Cabras, I felt that the

> creation of an MPA was [perceived more as a] danger by my fellow citizens, rather than as an opportunity. (Appendix, Interview, Group A)

> I was afraid of merging all those municipalities into the project of launching an MPA because they would certainly have the potential to influence us a lot in terms of timing [.] Every city council would have to deliberate and accept the same regulations as the neighbouring municipalities. [. . .] The past has shown us [. . .] how many years it takes to see such an MPA [fully established] [.] If there had been five or six different municipalities discussing the same things, nothing would have been accomplished. It is already difficult to align the governing majority and opposition around an issue within a single city council [.] That's why I wanted to introduce a proposal that didn't require us to share anything with the other municipalities within the same provincial territory. I believe that, if we had involved, for example, the municipalities of Riola, San Vero Milis, Oristano, at that time [. . .] today we would still be [at the beginning stages] with the creation of our MPA. (Appendix, Interview, Group A)

Apart from the aspects already identified, this all had the effect of excluding the policy formulation and decision phases of non-institutional partners and neighbouring municipalities who were potentially interested in creating the MPA. This prevented the establishment of the most extensive MPA in the Mediterranean Sea:

> In my capacity as the president of an environmental association, I must point out with regret that [assigning the management of the marine area solely to the Municipality of Cabras] represented a great limitation in terms of the environmental protection goal, because, with more municipalities involved, the area would have become one of the largest protected areas in the Mediterranean, with a great variety of natural assets, and a greater potential. (Appendix, Interview, Group E)

In wanting to play a leadership role within the dominant coalition, the Municipality of Cabras estimated that the Ministry of the Environment would provide vast economic resources when the MPA was established and that if the municipality could gain exclusive control of these resources, they could increase employment and attract structural investment within the same territory, including through EU funding:

> I was serving as the Councillor for the environment in the [Municipality of Cabras], and suddenly something very important happened. At that time, I often met at the headquarters of my political party with [this] friend of mine who had been elected in the district of Oristano [for the Regional council of Sardinia]. One day, talking about the creation of a Marine Area in Cabras, this friend said to me: "[. . .] Why don't we do something about it? Let's organize a trip to Rome to meet the Minister of Environment. Then you propose the boundaries to him on behalf of the city of Cabras, without waiting for anyone else" [.] The national law [. . .] already provided for the establishment of [. . .] more than 19 [MPAs]. [. . .] Then my friend added: "You must be careful, because if you don't do it this way, others will impose many constraints and restrictions on you which won't be good for your fishers, for your citizens, for a whole range of things. You should identify the MPA and its boundaries within the city council area. And this includes all the restrictions, [. . .] the no-go zones or [. . .] the partial restrictions for tourists in the area. I will organize everything in Rome at the Ministry of the environment for you, and you will only have to deal with the Cabras City council and local citizens in terms of discussing the creation of an MPA and preparing everything by starting the process [at the local level]". That's how it all started. (Appendix, Interview, Group B)

> For [the establishment of the] marine areas [...] a budget was provided [at a central level of government]. At that time, it was a lot of money for the Municipality of Cabras [...], and it would allow us to establish this marine area as if it were a start-up company. We would define the boundaries, organize it, buy everything that would be used for it. There would have been funds for everything [we needed]. (Appendix, Interview, Group A)

> [The ministerial financial and economic resources] were certainly important [in the decision to establish the MPA in the Municipality of Cabras only]. But rather than the resources of the Ministry of the environment, we were more interested in the resources that we could get from the EU. Because there were large resources available from the EU for the right sort of environmental projects. So, we thought it could be an asset from that point of view, more than the ministerial funds, even though in those years there was much money for the MPAs granted by the national government. (Appendix, Interview, Group A)

> [The economic and financial] resources [...] [coming] from the Italian government were abundant [...]. [The resources were necessary] [...] because you had to [create] infrastructures for the marine area [...] in order to have benefits, to be able to [...] make [...] workers accept the area, [by giving] them the possibility to be employed in other MPA-related activities as well. Because within a short [time], many fishers, about two hundred or three hundred of them, would not have been able to survive under the new rules and so then you had to [...] diversify [the activities of] these people into other activities, still related to the sea, such as recreational fishing [.] Without the economic resources, [establishing the MPA would have been impossible]; you needed to have the economic resources [to do it]. (Appendix, Interview, Group B)

> Of course, [the economic and financial resources] were vital in the discussion [about establishing the marine area in Cabras]. Keep in mind that we are talking about the late Eighties, so that was a time when there were many public resources [at stake]. And even the establishment of a marine area could make administrators think, and I think they were right at the time, that the marine area would be supported with substantial funding [both from the central government in Rome and the EU]. The [resource-related] aspect was [...] very convincing for the local administrators. Then, of course, we began to discuss also the possibility of protecting seawaters [from an environmental protection point of view] and promoting tourist initiatives [from a local development point of view] that could certainly be beneficial for the establishment of the marine area itself [and for the whole local community]. But I believe that this awareness built over time. (Appendix, Interview, Group A)

With these premises in mind, subsequent Cabras city councillors over four different municipal administrations maintained a constant engagement with the Ministry of the Environment in order to obtain direct and exclusive management of the MPA, with the publicly declared objective of safeguarding the local population and the interests of the community, especially the interests of fishers and tourist operators. It is clear from the interviews carried out that those political and social actors who were excluded from the dominant coalition while the policy was being formulated had faith in the role which the Municipality of Cabras proposed to undertake. The interest and pressure groups, as well as local public opinion, in fact, took for granted that the local administration referred to would be able to listen to, understand and respond to the needs of the population, thanks to their familiarity with the socio-economic problems of the area and their greater capacity to work with local social actors compared to the rigidity of the ministries and the Sardinia Region, as well as with greater democratic legitimacy compared to the environmental organisations:

> In the early Nineties, the Municipality of Cabras had applied [to the Ministry of the environment] for permission to manage the marine area. [...] We believed that it could be managed by the Municipality because it was more active than other [public bodies] within the area. The management of the area by the Municipality was aimed at representing the citizens and showing them that the whole MPA could become part of their heritage [.] The fact that the area could be managed [directly] by the Municipality [of Cabras] was really a very positive thing for us because, at least at that stage, the environmental protection policy was accepted by the population. I still think now that it was a good administrative choice for everyone involved. (Appendix, Interview, Group A)

> In our opinion as local policy-makers, the only way to make the marine area viable was to have it managed by the Municipality [of Cabras itself], because the Municipality is close to the citizens [or at least it is closer than other public institutions are] and, therefore, the Municipality knows their needs and what kind of development is right for its people. How could the Ministry of the environment manage something located so far away from it? (Appendix, Interview, Group A)

> We felt it was fairer to entrust the management to the Municipality [of Cabras] rather than to the Region of Sardinia [...] [because the Municipality is] the first contact point for the citizen, the first provider of services [.] The [...] public authority that is closest to the citizen is the Municipality [and] not any private association [...]. [We thought that associations] must serve as support, just like central and regional institutions. (Appendix, Interview, Group B)

But paradoxically, many interviewees highlight the scarce or no consideration given by the Municipality of Cabras as the selected commissioner to narrowly socio-economic factors linked to the activation of the marine area, compared, for instance, to environmental aspects. According to these same interviewees, this was an error in the formulation and decision phase, whose consequence became evident above all in the implementation phase of the policy, with the request for the revision of the regulations and zoning of the MPA, only 2 years after its approval, by various pressure and interest groups active in the local area:

> The MPA of Cabras was part of that first round of MPAs which were established in the late Nineties. The economic resources were not lacking at the time, and they were used to carry out [...] extensive studies from an environmental protection point of view. As a matter of fact, the social and economic studies about the area came later. This was something that we learned by doing, so to speak. That is, in the beginning, the studies for the establishment of marine areas focused mostly on naturalistic issues and [on naturalistic] elements to be safeguarded within the protected areas. (Appendix, Interview, Group B)

> When a territory that is trying to create opportunities for development instead suffers restrictions designed to protect its most sensitive and valuable environmental and landscape-related elements, this creates serious conflicts. That is why there was a popular uprising during the implementation of the [environmental protection] policy phase, an impetus that came from the local area, claiming that changing the framework of constraints developed for the marine area was not acceptable [to many parts of the community]. There was also a lack of engagement on the part of the law enforcement bodies, the Port authority, and other marine authorities, which avoided close monitoring and implementation of the

regulations. Indeed, the marine area [as it was designed theoretically] was impossible to enforce in reality, even from the perspective of public order. (Appendix, Interview, Group B)

More generally, we can note how the dominant coalition was made up, above all in the early phases of the policy (from first entry onto the agenda until the decision), solely by institutional actors (in order of importance: the Municipality of Cabras, the Ministry of the Environment, the Sardinia Region), and that the functions and responsibilities were distributed within a framework of interdependencies which excluded non-institutional figures, particularly environmental organisations, but also epistemic communities and scientific committees:

Most City councillors believed that it was necessary, in the creation of the MPA, to avoid at all costs it being controlled by the environmental associations. The people involved in the political life of Cabras considered the environmental associations to be fundamentalists who would oppose any tourist development on their coasts. Those were the years when we imagined that we could achieve tourism development through the construction of coastal villages, large infrastructures, so the idea was that the marine area, an instrument like that, could, in theory, have a certain influence on the development choices [.] The idea that it could end up in the hands of environmental associations for these reasons frightened many people. Hence, the commitment of the Municipality of Cabras to establishing this instrument on its own, preventing the environmentalists from managing it directly. However, the municipality itself was unable to understand the social and economic needs of the population and ended up taking care only of the environmental needs of the area. (Appendix, Interview, Group B)

[The Municipality of Cabras] had at his disposal pre-set models [for establishing the area] provided by the Ministry [. . .], take it or leave it. [. . .] So we didn't do any research on the local sea area to understand what the specific needs were, what the social and economic peculiarities of the area were. I don't know, and anyway, if this was done [. . .], it had no result whatsoever, meaning that the choices made were not developed based on the needs of the sea and, especially, [on the needs of] the population. (Appendix, Interview, Group A)

From the point of view of the distribution of functional responsibilities among actors in the coalition, it is useful to underline how the framework of interdependence does not seem completely clear to the actors involved, at least until the implementation phase of the policy. This is a function of the fact that the actors in the dominant coalition had difficulty in defining the nature, boundaries and functional limits of the marine area and consequently in defining who in practice had responsibility for managing the area:

[The approach to establishing the MPAs] was rather centralist [. . .], even though the Ministry [of the environment] did not have a structured relationship with the [local government and the other peripheral bodies]. At that time [I have to admit] there was not a great deal of willingness to have a protected area in Cabras either. The MPA was seen as a constraint, so after the first enthusiasm, there was a lot of mistrust about accepting it [and] or having a dialogue with all of the actors involved. [. . .] In a nutshell, I think [many stakeholders] were not involved in the preparatory phase. [The City council] was perhaps the only actor that governed the process, or at least it wanted to govern the process. (Appendix, Interview, Group A)

When I governed the Municipality of Cabras [after the national law was adopted], we started to make contact first with the Region [of Sardinia], although, about the MPAs [. . .], at first,

> we were not talking about that exactly [but we were talking about establishing] the 'Sinis – Mal di Ventre' natural park. Originally, it was to be a Regional Park, then it became an MPA. [. . .] The park would have been run by the Region and would also incorporate neighbouring land, that is, the land close to the marine area as well. [We didn't want that outcome and] we then reopened contact [and] resumed discussions with the Region which in those years had administrators who were also quite sensitive to the environmental protection issue. [Together we] started the process for [establishing] a protected area instead of a park. As a Municipality, we asked the Ministry of the environment to appoint us as the sole Selected commissioner [excluding everybody else]. (Appendix, Interview, Group A)

> The boundaries of the MPA set at the very beginning included only the territory of the Municipality of Cabras. At later stages, there was an attempt by the Province [of Oristano] and other municipalities in the area to join the MPA. But at that point, we were quite determined to have the Municipality of Cabras as the sole Selected commissioner. [For this reason] we immediately resumed contact with the Ministry of the environment, asking them to assign the MPA to us. The [Sardinia] Region [as well as the other municipalities] had to accept this. (Appendix, Interview, Group B)

The absence of or difficulty of interaction among institutional actors was linked by many interviewees to the inability of the Ministry of the Environment to effectively play a *super partes* role in the intermediation of interests or to foster cooperation among the political and administrative actors at different governmental levels (e.g. the municipalities and the Region of Sardinia) within a policy network that was intentionally built to be state-centric (Tebaldi and Calaresu 2009, 50–58):

> [The first] proposal for the marine area was worked out [. . .] on the basis of information [. . .] which was rather inaccurate, [I mean] not based on geophysical studies or anything like that, [but only] on [anecdotal] and personal knowledge [such as] 'things heard from the fishers', 'from the older adults' and so on. (Appendix, Interview, Group B)

> I wish I could have done something different, [. . .] but at that moment it was [. . .] too difficult, [. . .] it needed some more in-depth study. [Others were taking care of it] at the national level, but at that time, I was in a hurry [to establish the MPA at the local level], and so this seemed the most suitable solution. [. . .] In my capacity as a Mayor, I had little to choose from: [. . .] there were two technicians in the Municipality [of Cabras], one who was involved in urban planning and one who dealt with public works. [. . .] I told the urban planner [. . .] [to] take the car, [to go to] set the boundaries of the marine area, and then to discuss [that boundary] with the Ministry. It sounds [foolish] now, but that is exactly how it happened. (Appendix, Interview, Group A)

Moving on to a definition of the framework of interdependences relating to bureaucratic–administrative actors, it is possible to observe how the relations between bureaucracy and local government were dominated by clientelistic exchange rather than actual reciprocal feedback between executive and decision-making actors. In this case study, as theorised in the political science literature referred to, these contacts seemed, for the most part, to be focused not on 'the elaboration of policies and related decisions, but rather on more limited measures, sponsored for clientelistic or at least particularistic motives by political actors' (Guarnieri 1989, 227). The lack of technical–scientific study during the formulation and decision phases of the policy cycle seems to have had significant consequences in the implementation phase, particularly in relation to the definition of the framework of interdependence among actors in the dominant coalition.

It is only after the first years after the establishment of the MPA that the relationship between scientific research and the Municipality of Cabras as the selected commissioner tends to become constant in terms of time and effective in terms of quality. In fact, the reformulation carried out on the area's original establishing decree is based at least in part on the results of the studies commissioned by the research bodies in the years following the start of the policy. Despite this, the desire to bring about modifications to the original boundaries of the MPA seems to have been influenced above all by private stakeholders, as noted by our interviewees:

> As a local government, we had [relationships with scientific researchers], [...] [although it was] more like a liaison activity with a relevant social actor [of the community], than an official scientific assignment given to them by the Municipality [of Cabras]. We commissioned a whole series of research studies from the International marine centre to identify mitigation actions using specific environmental measures. For instance, the use of anchors in Posidonia seagrass areas. [Also] we set up floating anchorage buoys to [...] moor boats without requiring the use of anchors, which could have [damaged the seafloor]. We also took actions based on scientific research on sea urchins so that their harvesting would be regulated [.] There were many interventions of this kind, even if, to be quite honest, we needed [...] scientific research that would justify and validate the political decisions that the administration was promoting. Other [scientific research] was excluded from the table for the same reasons. (Appendix, Interview, Group B)

> [Studies which were considered] more urgent [were implemented in the MPA]. For example, the study on Posidonia seaweed allowed us to define where the mooring field should be established. The work with the International marine centre gave us information about the basic biology of sea urchins and how many sea urchins were present in the marine area, how many were being harvested, how many could be harvested. But [in the meantime] social issues were certainly taken into account, which determined the feasibility of policy decisions over scientific predictions. (Appendix, Interview, Group D)

To conclude, the level of interaction between the politico–administrative actors and the scientific research had poor results in the initial phases of the MPA but increased during the implementing phase of the policy. The experts '[...] who possess certain almost exclusive capabilities who are able, on the basis of those capabilities, [...] to suggest the most suitable means of achieving them' (Lanzalaco 1995, 118–119) did not have a determining role within the dominant coalition during the launch phase of this public policy. On the other hand, the selected commissioner, which was central in the initial phases, seems to have had to change its behaviour in the implementation phase by incorporating the various needs of the stakeholders and redesigning, at least in part, the map of interdependencies.

Degree of Congruence of Cognitive Maps Within the Dominant Coalition

The second dimension examined in terms of the institutionalisation of the governance system was the (b) degree of congruence of cognitive maps within the dominant coalition.

The empirical analysis we carried out took as its point of departure that the rationale for establishing MPAs in Italy is to maintain as a function of the State (and specifically the Ministry of the Environment) the safeguarding and protection of particular available areas identified as worthy of protection. This is because within Italian legislation '[...] based on arrangements set out in art. 25 of Law 979/1982, articles 18 and 19 of Law 394/1991 and constitutional jurisprudence, responsibility for domestic waters as well as the establishment of reserves of national interest is a function of the State' (Marino 2011, 74). This has meant that in the typical case study we are interested in, the Ministry of the Environment has maintained a role of absolute centrality in the construction and control of the cognitive maps within the coalition, both in terms of the process of establishing the protected area and in terms of all the phases in its management.

What emerges from the interviews carried out is that the substantial absence of the principle of subsidiarity in relation to the Municipality of Cabras had significant effects on the degree of congruence of the cognitive maps within the dominant coalition. The Municipality of Cabras, although formally the selected commissioner for the area, was in practice substantially relegated to responsibilities for management and planning of the safeguarding operations and for guaranteeing the ordinary functions of the MPA. It was compulsory for all of the most important legal documents (e.g. the implementation and organisational regulations for the MPA) to be approved by Decree of the Ministry of the Environment. The result of this hierarchical top–down-type process was a significant slowing down in the implementation of changes requested by the municipality, the associations, the interest groups, and the local population during the policy reformulation and implementation phase, reducing the ability to propose and implement changes (e.g. by periodically 'rotating' type A zone areas in order to ensure full protection). From this point of view, the interviewees highlight, in particular about cognitive maps, the lack of distinction between technical and political decision-making, which caused an increased bureaucratic burden, and a lack of communication between the various actors in the daily management of the MPA:

> It took 10 years to change [the law], and since 2006, enforcement regulations [for the MPA] have still not been approved. [It is a political issue]: [...] the Minister has to think about overall political strategies and cannot worry about the day-to-day management of the MPA. As a local administration, we paid the price for this top-down approach that gives the central government veto power over every decision we make [at the local level]. In my opinion, regulation of the MPA has to be addressed and evaluated solely from a technical perspective, not a political one. Otherwise, even the administrative choices are compromised by the presence of the national policy-makers [.] There is a disproportionate interference of politics in the bureaucracy, there is no doubt about that, [plus] a basic distrust in those who have to carry out technical-administrative activities, and all of this is a limitation [for the implementation phase of the policy]. (Appendix, Interview, Group A)

> There are bureaucratic problems [...] because unfortunately Italy is made up of many agencies and offices linked to laws and rules that are sometimes not clear. There are always many different competencies and people expressing an opinion from different perspectives. (Appendix, Interview, Group D)

On this front, a further strongly negative element imposed by the legal norms was that it was impossible for marine reserves to appoint their staff (Art. 8, Law 179/2002), given that management roles can only be assigned from already available staff:

> The laws in Italy forbid the MPAs from equipping themselves with their own workforce, so they can only count on already available staff [.] The [. . .] law [. . .] clearly states that the staff must be provided by the Selected commissioner chosen by the Ministry. In our case, that means that the Ministry not only does not provide for new staff to be used in the MPA but does not even allow the Municipality [of Cabras] to hire new people. You [the Municipality of Cabras] want to run an MPA? All right, you have to use the staff you already have at your disposal. The Ministry will cover only the MPA Director's expenses. Everything else must be provided or hired by the Municipality. (Appendix, Interview, Group A)

Degree of Structuring and Impermeability of Gatekeeping

The third dimension in the institutionalisation of the governance system relates to the (c) degree of structuring and impermeability of gatekeeping, particularly in relation to the actors who are able to control, at least at the major nodal points, the political, legal and economic–financial resources of the decision-making process.

As far as legal resources go, one important nodal point in this typical case study relates to the introduction, following the reception of the instituting decree for the area, of the implementation regulations concerning permitted and prohibited activities within the area. In fact, as already highlighted, about 2 years after the area was declared, it underwent some changes to boundaries and zoning, which led to a significant delay in the creation of the regulations cited, which were necessary for the efficient management of all activities (from fishing to tourism) considered possible within the reserve. In the absence of the implementation regulations, the daily management of the MPA was carried out for a long time via municipal ordinances (Calaresu 2017) which—above all in the summer season—were issued urgently as the case required by the mayor of the Municipality of Cabras, in his role as a governmental officer. Because of this, the rules of the game were debated each time and in each case by the actors involved, with a consequent reduction in the degree of structuring and impermeability of gatekeeping. The same mayor who filled the role of MPA president in those years highlights this aspect as extremely significant in terms of the degree of structuring and impermeability of gatekeeping:

> [To manage the MPA] in the absence of day-to day regulations, as a Mayor, I used to issue administrative orders which I discussed in advance with the citizens and their social and political representatives. In this way, I was pretty sure that the orders would be accepted and supported by almost everyone. (Appendix, Interview, Group A)

The pressure and interest groups, committees, political parties, representatives of unions and environmental associations—as we have seen almost totally excluded from the dominant coalition in the policy formulation and decision phase—acquired in this way a more central role in the policy reformulation and implementation phase of the policy (1998–2004):

> [The use of administrative orders] was a compromise linked to the fact that the first version of the policy was strongly criticized. This was because the social actors had not been involved from the beginning, and this situation led even to problems of public order between the fishers and the law enforcement agencies. (Appendix, Interview, Group A)

> The first boundaries of the MPA were the result of the lack of experience of the local administration [in Cabras] and its lack of ability to predict what would happen to the words that they wrote on the paper, which were often written solely on the basis of environmental protection concerns, and did not consider economic aspects. Once you apply the rules you have made to reality, you have to face the clash of those rules with the needs of the recreational or professional fishing sectors, the tourism sector, and so on. So, the [contact between political decision making and reality] led to a need for the area to be re-evaluated, also [...] because of the fact that, to be honest, the first local administration in Cabras [...] did not last very long in power compared to the time needed for the implementation of the protected area [.] In fact, shortly after the MPA was established, the [municipal] administration crumbled [and] new elections were called]. (Appendix, Interview, Group E)

> [The review of MPA boundaries took place out of] a social, rather than a scientific need. We realized that we could not ban too many things inside the MPA. With the administrative orders and the ensuing regulations, an attempt was made to balance environmental protection and tourist development, creating conditions of accessibility and use for the whole area, instead of prohibitions only. (Appendix, Interview, Group A)

> Initially, [in the first MPA boundaries] [...] there was a full reserve protected area, but the fishers asked us to deregulate that special protected area because that is one of the few closed natural harbours on our coast where boats can find shelter in case of strong winds. If the MPA had prevented them from anchoring there, there would not have been anywhere for a boater to take shelter in case of a storm [for many miles], and that would have put an end to recreational boating and professional fishing because of the MPA. (Appendix, Interview, Group B)

> I am convinced that maintaining the initial boundaries of the MPA would have led to [its] closure, that is, to [its] rejection by the whole community within the local area. If we had not been able to redefine the boundaries and regulate them, we probably would not have had the political strength to continue the policy of environmental protection. We would have encountered an overwhelming social and political opposition, that would have forced us to end the protected area. (Appendix, Interview, Group A)

In the implementation phase, barriers to access were lowered compared to the preceding phases because the non-institutional groups gained greater indirect control of resources through their intervention in the decision-making process. In particular, the municipal administration was forced in the reformulation phase to allow into the decision-making arena, groups who were external to the dominant coalition and providers of various resources, for example, political, economic, informational or technical–bureaucratic. This was part of an attempt to get the local population and interested social groups to accept restrictive measures that constrained activities (professional or leisure) which prior to the establishment of the MPA were not subject to regulation or were at least lightly regulated. This was all motivated by the danger of compromising or possibly having to abandon the entire MPA project because of the political–social costs associated with it:

> The re-evaluation of the MPA's boundaries was a compromise between the different stakeholders like the fishers, the boaters, the marine tourist operators, the divers, etc. At a certain point, it was necessary for the local government in Cabras to try to mediate between these different needs [.] It was impossible [to maintain] a boundary based solely on the desire to preserve a certain natural ecosystem type. We couldn't say that no one could go to Mal di Ventre [anymore. . .] because that was the area inside the MPA with the greatest biodiversity [.] You just can't remove a landing point for boats like [. . .] Mal di Ventre [. . .] which may also be the only area of the MPA where tourists can go diving. (Appendix, Interview, Group B)

> The Municipality of Cabras] listened to us fishers a little [. . .] [during] the [formulation] phase and a little more in [the] [implementation] phase [.] The only changes that we achieved by protesting were [. . .] some reductions in the total extent of the 'full reserve' area of the MPA. [The policymakers] thought that we would be satisfied with reducing the size of the area. [But] we didn't want that. What we wanted [actually] was the full reserve areas of the MPA to be closed for a certain period of time and to be reopened for fishing afterwards. (Appendix, Interview, Group E)

Another significant element with the control of major politico–legal resource nodes was the negotiations during the policy reformulation phase between the Municipality of Cabras, the Sardinia Region and the Ministry of the Environment. In these negotiations, the municipality, even though it had been given formal responsibility for managing the area, was subject in practice to the two bodies superior to it in the hierarchy; this was because the municipality lacked the necessary legal and political resources to enforce its own choices but also because it was subject to the interests of narrow segments within its population:

> [In terms of the change to the MPA boundaries] all I can say is that [in the formulation phase of the policy] the Municipality of Cabras listened to the fishers who were most caught up in the past, even if they represented only a minority. Those fishers were only interested in their immediate day-to-day earnings. For a while, we as policy-makers maintained the MPA areas we thought would be useful for project funding. Once some of our projects did not get financed, the fishers who were against the MPA had an easy time saying, "the reserve areas you have taken away from us are too large" or "you are ruining our jobs". So, we needed to reach a compromise, which was the result of two wrong things. (Appendix, Interview, Group B)

Finally, the relative scarcity of politico–legal resources (e.g. the shorter term of municipal mayors and administrations compared to the longevity of ministerial technical officers, the lower degree of training and specialisation of the municipal technical officers compared to those in the region or Ministry) led in the later phases of the policy to a weakening of the ability of the municipal administration of Cabras to control major nodal points in the process, with a consequent further lowering—compared to what was noted previously in relation to the external pressures on the coalition—of the degree of structuring and impermeability of gatekeeping:

> [The Ministry] was also very open [to the proposals submitted by local authorities about the regulation of the MPA after its adoption]. The problem was that there was a very complex bureaucratic and institutional process that was not very receptive between the Ministry and the local administrations. [The whole process] was full of documents, memos, bureaucratic contacts. Making a telephone call was a luxury. So [. . .] there was only an exchange of

> documents and, in the absence of an answer, since local administrations often fell [or] the mayor changed, [...] the Ministry went on with its own procedures because it had not [...] received an opinion from the local administration, and the law said that the Ministry simply needed to communicate, not to wait for a response. So, in terms of whether there was any benefit for the Municipality coming out of all this process, let's say that it was quite irrelevant. This resulted in much resentment among the local administration [and] local citizens. Local policy-makers used to say "But we didn't even have time to reply", and the Ministry would say in response: "I don't care about that, the [only] important thing according to the law is that I communicate to you that I am doing such and such. If your predecessor hasn't replied, I really don't care". That was essentially the ministerial approach [at the local level]. (Appendix, Interview, Group D)

> The competence for sea areas lies with the State, [...] the municipalities have no jurisdiction over the sea. In some ways, this empowered the Ministry to operate in the absence of involvement with or sharing with the local administration. (Appendix, Interview, Group A)

From this point of view, it is possible to confirm on an empirical level what is stated within the political science literature referred to in terms of the unsatisfactory performance of bureaucracies, of the frequent suboptimal solutions and of the continuous adjustments between rules and politico-social demands in the course of policy implementation and reformulation, as characteristics common to all regulatory processes (Lizzi 1995, 114–115), processes that are usually characterised by innovation, interdependence and complexity. With regard to the control of economic–financial resources too, even though, as noted previously, resource acquisition may have represented one of the principal motivations around which the direct and exclusive action of the Municipality of Cabras was based, according to the interviewees, this municipal actor showed clear signs of failure and inadequacy in the control of the major nodal points of the process in the later policy phases. Statements by insiders allow us, in fact, to note that despite the considerable flow of resources obtained, thanks to the establishment of the MPA, which were intended for both the day-to-day management and the development of activities and/or creation of services or structures (Marino 2011, 214), many of the projects—in particular those of an infrastructural character—were never realised or fully activated:

> I believe that financial resources were allocated and actually provided to the Municipality by signing the 1998 Agreement. I believe that, on that occasion, the Municipality of Cabras received a lot of money [...] from the Ministry and from the Region of Sardinia. (Appendix, Interview, Group A)

> Apart from fishing, we thought that other resources could be obtained for the marine area. For example, we [...] had also prepared the preliminary plans for the use of existing facilities and the use of other facilities to be built within the MPA boundaries. The museum, a multipurpose centre [...] was to become the Museum of the Three Seas. It could have been the beginning of a whole local development plan. In San Giovanni, where the Cove is located, there is [a] facility [...] [that] we had already requested and we were planning to relocate the aquarium there, to make sure that people came to Cabras for tourism. (Appendix, Interview, Group B)

> We had planned the construction of several marinas, one in San Giovanni, another in Seu [...] and another one in Porte Suedda [...] and another in Mal di Ventre. We wanted to create a [network] between these marinas [...] so that those who arrived in Sardinia could spend some days in Cabras. But this big project around the use of the MPA [...] remained a dead letter, nothing was done. (Appendix, Interview, Group B)

> The idea was to create a scientific and technical structure that would involve local professionals, giving an opportunity for the kids to deepen their environmental studies, then to create [in one generation] the professional skills that were lacking in Cabras [for running the MPA]. I must say that [initially] the intention of the Ministry was to create a Mediterranean marine environmental centre in Cabras [.] They relied a lot on our marine area because it was in the centre of the Mediterranean. For them, our MPA was [...] the perfect place to focus [...] efforts to create many MPAs in Italy and around the Mediterranean basin. For us, one very interesting project was to encourage young people to follow this kind of specialisation and to promote the environment, tourism, fishing. For us, it was important. We relied on it. But we were not able to realize any [of that] because after signing the establishment of the area, my council ended its mandate, and the following administration was not able or willing to implement the project. (Appendix, Interview, Group A)

Even the projects managed by the Municipality of Cabras in collaboration with the EU, at the suggestion of and with the participation of the environmental associations (including, for example, *Legambiente*, WWF), had the same negative outcome, as a consequence of the incapacity of the selected commissioner for the area to control the nodal points of the decision-making process:

> The Pauli and Sai Oasis was also established, on the initiative [...] of the WWF, which launched the idea of having [European] money to build and safeguard the Cabras wetlands. So, the WWF [...] presented the project, and the Municipality of Cabras was favourable [and] contributed to it [economically] [...], with over 250 thousand euros. [This resulted in] the improvement of the access roads, parking, signs, [...] the construction of two huts within the reserved area [.] But within a few months, the huts no longer existed. [The project] died. Now everything is back as it was before, there is nothing left. The area became covered in brambles and weeds. Cabras has gone back in terms of the environment and its protection, but also from the point of view of local development, it has gone back many, many years. Everything that we built positively to do with the protected area up to the Nineties [...] now no longer exists. (Appendix, Interview, Group E)

The interviews also show that in the policy reformulation and implementation phase, the Municipality of Cabras acted following purely politico–electoral objectives (e.g. victory in a municipal election campaign), as well as with objectives dictated by the necessity to lessen pressure on the dominant coalition (in particular from boat operators and fishers), instead of in accordance with long-term plans for local development and environmental protection, with immediate actions being focused on increasing social acceptance of the MPA:

> [The MPA could have been] a resource for fishing as well because in those years there were great funding opportunities at the EU level for those who worked in fishing. When the local administration changed, those who won the elections won by shooting point-blank at the MPA, and none of those big projects was ever carried out or implemented. Not even one. Indeed, I believe that the MPA was used to do things that maybe the administration could not do, [such as creating political clients]. For this reason, too, there was much less control over the marine area. In a sense, over the years, the MPA has lost [its connection to] the real

reason why it was created, which was to stimulate the social and tourist development to the whole area. (Appendix, Interview, Group E)

The population's demands are very utilitarian [...] they are never based on cultural issues. The questions [raised by people] are [something like] this: "What does the MPA give me? Does it suit me? How much do I get out of it?". That is pretty much [the] common thinking [.] We were an agency that someone said [would only exist to] distribute money and funding. But I have to say that other institutions saw it that way too, other institutions such as other municipalities [within the territory] of the Province of Oristano. They were constantly asking for money from us. In practice, as the Selected commissioner of the MPA, we were seen as an agency that distributes funding [.] It is therefore still necessary to make the connection between environmental protection and economic advantage, and cultural advantage, because at the moment there is no such awareness. (Appendix, Interview, Group D)

In conclusion, the control of resources in the formulation and decision phases was governed at the local level by institutional-type public bodies such as the municipality and region, but in the reformulation and implementation phase, these same bodies were not able to exercise their functions in a manner which was completely autonomous of sectoral and micro-sectoral interests, within a system which tended to be relatively permeable. The particular critique noted in relation to this was, above all, the distributive nature of politics, which allowed actors external to the dominant coalition easy access to the resources at stake.

4.4.1.2 The Quality of the Environmental Protection Policy

Degree of Responsiveness

Turning to an analysis of the quality of the urban regeneration policy, the first dimension examined is the (a) degree of responsiveness, which can be detected by comparing the expectations and objectives of the intervention shared by citizens as well as the local ruling political class, with the actual results achieved by the policy.

In this case study, the interviewees were practically unanimous in highlighting that the tourist sector, which was supposed to constitute a strategic sector following the establishment of the MPA, did not manage to gain a level of development adequate to expectations. The creation of the MPA did not produce tourism opportunities capable of fully valorising the ecological, archaeological and cultural heritage of the area. This was mainly because of chronic deficiencies at the infrastructural and local entrepreneurial level, which had never been resolved. From this point of view, the interviewees highlight a low level of policy responsiveness as a driver of territorial and infrastructural development:

Thanks to the marine area, there have been many environmental restoration and territorial infrastructure development initiatives, but we continue to suffer a basic problem: the lack of entrepreneurial skills, the low inclination for investments and the lack of capital. For example, the cooperatives [involved in the marine projects] went ahead only as long as they had orders from the government to go ahead. When they had to enter the open market, [the difficulties began] [. . .]. So, we invested a lot there, we developed a lot, and many services were offered, [. . .], but when the cooperatives did not have contracts awarded directly, they struggled to stay in the market. (Appendix, Interview, Group A)

In Cabras, there is very little entrepreneurial mind-set, and unfortunately, contrary to expectations, the establishment of the marine area failed to solve the problem. When you don't believe in a thing, the input doesn't come, and everything stops [.] In Cabras, people [. . .] [have no] entrepreneurial skills [. . .] and, therefore, the local administration had to push more to compete with other places. (Appendix, Interview, Group E)

Secondly, even the objective of protecting and valorising the biological and geomorphological resources of the area and the replenishment of fish stock in the MPA (Art. 3, letter *b,* Ministerial Decree 1999) does not seem to have been achieved by the policy under consideration. On the basis of the interviews, the main cause for the failure to achieve this objective lies in the provision of regulations which were too rigid during the formulation phase of the policy, and in the later absence/lack of enforcement of these regulations in the implementation phase:

The environmental objective of an MPA is to encourage the restocking of fishing and biological resources [. . .], however, the marine areas over time have also become laboratories for sustainable development, and areas where you could experiment with integrated management models for the coastal strip [.] As far as the achievement of those objectives is concerned, after the establishment of the marine area, there was no scientific evidence to prove the effective restocking of fish [.] The problem for Sinis [. . .] but also for many other marine areas in Italy, is the lack of well-organized monitoring of those areas. (Appendix, Interview, Group E)

As for the objective of the area in terms of local development, I don't believe we have achieved the desired objectives. I mean, on the one hand, we paved the way for the development of this area [. . .] with the commissioning of more innovative coastal management models [. . .], but on the other hand, [there] was no major cultural development, and sectors such as fishing and tourism have lagged behind the envisaged goals. (Appendix, Interview, Group B)

In particular, the interviewees agree on how little the objective of ensuring sustainable fishing in the marine area has been achieved and on the policy's poor results in terms of improving fish stocks, as a consequence of the scarcity of controls and the lack of application of sanctions provided for violating rules and regulations:

[The Municipality of Cabras] as the Selected commissioner [. . .] during the implementation phase began to question every rule [such as, for example, the period and locations for fishing]. But in all this [I think] that the surveillance of the area was never adequate because no one went to check that the rules were really being applied, or if [someone] checked, they turned a blind eye, for example about how much [fishing] gear was taken on board and whether this amount corresponded to the provisions of the MPA [.] In a nutshell, the rules were not always enforced by those responsible for ensuring compliance. (Appendix, Interview, Group E)

If the objectives set by the environmental protection policy were to be achieved, fishers had to be involved [in the first stage]. During my administration, we tried to [. . .] change [things] in that direction, but we only succeeded in part. Especially because from the very beginning, we saw the lack of investment in law enforcement for such a large marine area. If you entrust it only to the bodies of the state, it required energies that the Carabinieri and other enforcement agencies did not have [.] This would have required greater commitment by the fishers. On the one hand, so that they would be more willing to respect the rules voluntarily, and on the other hand, so that they would help the police forces in surveilling the area. (Appendix, Interview, Group A)

Degree of Cooperation

The topic of controls within the MPA can be used to examine the second dimension of the quality of environmental protection policy, that is, the (b) degree of cooperation. The laws referred to (in particular Law 979/1982 and Law 394/1991) delegate surveillance activities within the MPA to the port authority, with the support of other competent forces of the State (forest corps, *Arma dei Carabinieri*, national police, *Guardia di Finanza*) and the forces of the selected commissioner delegated to manage the area (in this case, the Municipality of Cabras). According to provisions established during the policy formulation and decision phase, surveillance of the area was to be carried out in coordination with the staff of the selected commissioner who had responsibility for service activities within the protected area. In the implementation phase, however, the relevant actors noted a low degree of cooperation between the actors involved, with the Municipality of Cabras revealing itself incapable of controlling this process with its resources or coordinating the operations of external entities responsible for law enforcement. These external entities complained about their part of a chronic lack of personnel and available resources to carry out the tasks assigned to them:

There was no adequate law enforcement of the protected area: […] it was too occasional. [As a matter of fact] the Port authority and the Forestry corps were independent of the Municipality of Cabras as the Selected commissioner, so we were always on edge. We almost asked favours […] or at least we had to give [the police forces] support in terms of resources, including economic resources. For example, as a Mayor, I paid their summer meals because they had no money for that [.] There is always a gap between what the law says and what people can really do. (Appendix, Interview, Group A)

[The MPA Selected commissioner] has not criminal or judicial police duties. Those duties fell to [other police forces] such as the Port authority. It was up to them to [carry out surveillance of the area] and unfortunately, what they could not do for their own reasons, they came and asked us to do, in our capacity as the managing authority. (Appendix, Interview, Group A).

Any problem to do with the sea and which before the MPA was clearly the responsibility of others, since the creation of the MPA has become our problem, and this is [a] paradox [because] people seem to forget that there was someone who had to do these things before we existed. If a Port authority has to travel 10 miles […] to get to [our MPA in case of trouble], what kind of surveillance can they do? […] They can't shift blame for this to the [Director of the MPA]. These kinds of issues exist because the MPA was created on paper, without thinking about any logistical solutions for the implementation phase of the policy [.] We were not negligent, we were […] powerless, [and] we were being blamed for failure or inefficiency in things that should have been the responsibility of others, such as law enforcement within the area. (Appendix, Interview, Group D)

You can make all the rules you want, [. . .] but if there is no one enforcing them, then you have a paradox. If an area is not monitored, the first result is ineffectiveness. [The monitoring] is done by the police, but within their timeframe and to the best of their ability [.] So this is a fundamental issue that the establishment of the MPA has not been able to solve. If I want more surveillance as a Mayor, I have to pay the meals for the police forces [. . .] because they have no more money for functions that they should be performing. (Appendix, Interview, Group A)

Despite the constant communication between the actors of the dominant coalition (with the addition in the policy implementation phase of the province, the Oristano Prefecture and national law enforcement officers in the area), the problem of control and surveillance of the MPA was insuperable, just as it was in the formulation phase, with the municipality as the selected commissioner unable to carry out surveillance functions on its own or to manage/coordinate actors superordinate in the hierarchy and functionally independent (Calaresu and Tebaldi 2015). This low degree of cooperation resulted in widespread uncertainty about the validity of the MPA regulations and their application, as well as about the administration of penalties by responsible authorities against those who violated these regulations and a tendency towards conflict between the parties involved (Calaresu and Selmini 2017), both on the public–private (e.g. between selected commissioner and fishers), private–private (e.g. between fishers environmental associations and residents) and public–public axes (e.g. between selected commissioner and port authority):

I was in continuous discussion with all the economic and productive forces in the local area and with all levels of government, such as the Municipality of Cabras, the Province of Oristano, the Region of Sardinia, and the Ministry of the environment, including the law enforcement authorities, such as the Port authority, the National police force, the Carabinieri [.] It was not always easy because one of the main problems with the MPA, in my opinion, was that the law provides that the Selected commissioner does not also have surveillance powers over the area. The enforcement of the law and regulations are entrusted to other actors who are not hierarchically and functionally subordinate to [the municipality as the managing authority], but to the State [.] We did not have the power to intimidate or to order or to define guard shifts, operations, the intensity of presence in the MPA and so on. This was a very big limitation in the overall system. The operators of the MPA cannot intervene when people call us to notify a violation. I did an in-depth study with the then prefect to establish the Park Guards. They gave us an authorization that did not give us the power to impose sanctions or use prosecution to enforce the law effectively. Basically, if one violates the rules and a violation requires a sanction, that can be effective; but if one breaks the rules and the violation does not result in a sanction, you have a useless tool. And if you must involve third parties, the response times are those they choose, not those you require from them. At that point, it is very clear that the marine protected area is an ineffective tool. (Appendix, Interview, Group D)

When you see that, for years, the rules laid down on paper have not been enforced, there are consequences [.] For example, I think only two fines have been issued for trawling since the establishment of the marine area, and we went several times to complain to the Port authority, as small-scale fisheries because it was a scandal that perhaps a small boat was stopped because it did not comply with flagging rules, and then when there were much more

> severe violations, but those were not sanctioned. Yet the impact of small-scale fisheries is much less harmful than trawling. Since patrol boats did not go out at night, trawlers did what [they wanted]. So, you can understand that it is unlikely that you are going to be too enthusiastic about the technicalities of the regulations when you see that, in substance, they are a hoax. In short, the greatest outrage you can do to a marine area is having people trawling. So, the great pity for the marine area is exactly that: it is an empty shell with beautiful signs, but with very little impact on local development. (Appendix, Interview, Group A)

Degree of Functional Congruence

As far as the third dimension of the quality of policy goes, that is, the (c) degree of functional congruence, with particular regard to the capacity/incapacity to calibrate and steer programmed actions towards specific social target groups, we can turn our attention to two different categories of actors towards whom the environmental protection policy was directed.

On the one hand, a group of actors directly involved in the implementation of the MPA regulations in that they were employed in a profession linked to the exploitation of the marine environment and its resources (in particular fishing and tourism); on the other, another group of actors was indirectly involved in the implementation of the regulations and, despite not belonging to any of the preceding categories, was considered a target of public policy in the sense those actors might make use of the area (e.g. tourists or local residents). As far as the first category goes, it is possible to note that apart from some exceptions, the policy was not able to modify the conduct of these actors. Instead, the imposition of regulations, associated with a lack of surveillance in their implementation, seems to have generated new conflicts and non-virtuous behaviours characterised by repeated zero-sum game between the parties. In particular, concerning the relationship between the fishers and the regulations imposed within the MPA, it is possible to see in our interviews that:

> In Cabras, [...] the MPA [...] has certainly generated many opportunities. [...] The fishers' awareness that proper management of the coastline could result in more opportunities for everyone is widespread, which unfortunately, does not mean that everyone complies with the law. On the contrary, not everyone complies with the rules because one says, "if I follow the rules, I do not bring home what I need to live". So, they don't comply, and so I don't either. And here the circle becomes vicious, and it is like that most of the times. (Appendix, Interview, Group E)

Specifically, it is the local administrators who allow us to clarify the conflictual relationship between the policy and its target social groups and highlight a low level of functional congruence within the policy itself, at least in relation to the first category identified:

> When our [municipal] administration came to an end, the consortium of fishers had increased the turnover of the wetlands, I would say from something like 100 million to several billions of euros. Since they were aligned [politically speaking] with us, when the other side got in, they changed the governing board of the consortium, and the wetland again fell victim to all those who went there to steal. This [...] [just gives you an example of how]

amongst the fishers as well, there was a faction who did not want rules, did not want controls and wanted everything to be left like the Wild West. (Appendix, Interview, Group A)

Since the establishment of the MPA, fishing has changed for the worse in the sense that [for example] the amount of fishing tackle carried on board by fishers has gradually increased, although the regulations were seeking the opposite. The tackle issue is not insignificant because it is closely linked to certain types of fishing that were considered harmful to the environment and were intended to be banned from the protected area. There are a whole series of situations where there is still a very strong sense of individualism. Fishers would like more vigilance over others and a little less vigilance over themselves. (Appendix, Interview, Group D)

[Environmental education] would have been useful because setting new rules wasn't enough to get fishers to understand the real objectives of the policy, apart from limiting their activities. (Appendix, Interview, Group E)

[The MPA] created a major innovation regarding the governance of fishing. After the MPA was established, there was a method of evaluation that was not there before. [In the past, to catch] fish, [all one had to do was] lay nets, without following any rules and not worrying about the fact that any overused resource will eventually run out. (Appendix, Interview, Group B)

The empirical proof provided above allows us to highlight an important aspect of a cultural nature, concerning the level of functional congruence of the policy. In fact, the marine area was established in a town, Cabras, which in the past had experienced strong social tensions and conflicts precisely with the management of the sea and surrounding territory. For example, between the 1950s and 1970s, the fishers of Cabras were leading actors in a conflict with the local administration designed to obtain public management of the neighbouring wetland.[3] The transition which led to the Cabras wetland coming under public ownership coincided with the '[. . .] resurgence of that individualism typical of lagoon dwellers which in many cases has come to assume anarchistic tendencies' (Manca Cossu 1990, 18). The establishment of the MPA was considered by this category of actors in part as an aggravated and deleterious return to the restrictions of the past, with the outbreak of strong tensions in the implementation phase of the policy between the fishers and the governmental bodies (both at the local, provincial and regional levels), with the sole exception of small-scale fishers who were to some extent advantaged by the new rules.

In terms of the actors who were indirectly involved in the establishment of the reserve, considered target groups for the policy in the sense that they might use the MPA (in particular, tourists or just local residents), one of the objects provided in the ministerial decrees was the involvement of citizens through 'the diffusion of knowledge about the ecology and biology of marine and coastal environments, of the marine protected area and the special environmental and geomorphological characteristics of the area' (Art. 3, letter *c*, Ministerial Decree 1997) or 'the creation of educational programmes to improve knowledge in the field of marine ecology and

[3]For further information on these historical–social topics, see Manca Cossu (1990).

biology' (Art. 3, letter *d*, Ministerial Decree 1997). The creation of 'environmental knowledge' therefore represented an aspect considered by the policy-makers as very significant in guaranteeing the more general objective of safeguarding the protected environment. A first consideration regarding the behaviour of this second category of actors is provided by the two mayors who during their terms of office were also chairs of the MPA of Sinis peninsula – Mal di Ventre island and who give a positive view of the behavioural changes consistent with policy provisions brought about among citizens and tourists using the area (and more favourable in terms of behavioural changes brought about among the category of actors analysed above):

> I believe that after several years, even those against the establishment of the area acknowledged that setting up the MPA was useful. I think that many people have gained greater environmental awareness following its establishment [.] I have to say [that] as a matter of fact, the general feeling towards the protection of our sea has increased greatly, compared to other locations where such an area does not exist [.] Let's say there is now a greater awareness about the protection of the sea. So, in this way, the MPA was a positive thing. It resulted in a greater environmental awareness compared to the past. (Appendix, Interview, Group A)

> The municipal administration prepared itself by sharing its experience with the citizens [.] I remember that on the beach, a trivial example, if you saw someone smoking and smoking was prohibited, you reproached them. Even tourists called the police on such occasions; [. . .] everyone was surveilling the area, even if someone left a piece of paper or threw a piece of paper in the sea. There were boating associations involved in every initiative that had created this system of mutual coexistence, fostering the fundamental right to share the marine area. It is a social fact that citizens were involved in the management of the area. (Appendix, Interview, Group B)

From this point of view, the dominant coalition thus seems to have partially achieved the objectives identified in the regulatory phase by guaranteeing a medium to high functional congruence with the policy established in the regulatory phase, compared with a medium to a low level in relation to the category of actors employed in a profession directly linked to the exploitation of the marine environment and its resources:

> We carried out a big advertising campaign [to] raise awareness about the removal of sand from beaches [.] The goal [. . .] [was to focus on increasing] the sense that each one of us who is on the beach is the defender, the caretaker and the watchman over our neighbour on the beach who wants to break the rules. And this concept has indirectly been accepted, today, residents and non-residents alike have understood the message. And I can say that from the point of view of environmental protection, the MPA has been quite successful, for instance, one success has been stopping people from throwing away cigarette butts on the beach or the sea [.] Citizens' awareness of our environmental protection policies has improved. Progress has been made on simple things that are now common sense, and we no longer pay attention to them. But it has taken time to eradicate certain misbehaviours because we had to work on people's consciousness and people's consciousness is not naturally inclined to change. (Appendix, Interview, Group A)

Table 4.2 Overall evaluation of the level of institutionalisation of the governance system in Sinis peninsula – Mal di Ventre island MPA

Analytical dimensions of the institutionalisation of the governance system	**Overall evaluation in Sinis peninsula – Mal di Ventre island MPA**
(a) Degree of functional specialisation and interdependence maps among the actors of the dominant coalition	Medium/low
(b) Degree of congruence of cognitive maps within the dominant coalition	Low
(c) Degree of structuring and impermeability of gatekeeping	Medium/low
Level of institutionalisation of the governance system	
Medium/low	

> We let the population know about the marine protected area many times, we established awareness training in the schools. We organised public initiatives around certain themes, displays like "the marine area festival", fun events, cultural events. Some people understood the importance of these events, some less so, and others with no interest in this kind of public issues, others who see them as a huge opportunity to take part in the field of environmental action or tourism so that environmental excellence encourages certain types of experiences and helps you to build awareness. There are others though, who experience this more through fishing, whether as a hobby or profession and who have problems with the fact that they have to show greater care and can't go to some places, can't use more than three hooks or can't go there by car. So, the fact is that each of us individually may be receptive if we are sensitive to certain issues and absolutely non-receptive if we are not interested. But there's also the middle ground. (Appendix, Interview, Group B)

4.5 Overall Evaluations of the Typical Case Study

4.5.1 Evaluation of the Institutionalisation of the Governance System

In the case under analysis, as shown in the table below (Table 4.2), the first analytical dimension of the independent variable—that is, the (a) degree of functional specialisation and interdependence maps among the actors of the dominant coalition—can be considered as *medium/low*. This is because the map of interdependences is not sufficiently clear in terms of the distribution of functional competencies among the actors, even though during the first phases of the policy, the dominant coalition was made up only of actors associated with public institutions (in particular, the Municipality of Cabras, the Sardinia Region and the Ministry of the Environment), who distributed functions and responsibilities within the interdependence map in a way which excluded non-institutional actors. The reason for this is that the most significant actors in the coalition had difficulty defining the nature, boundaries and functional limits of the MPA and consequently determining who in practice had responsibility for managing them. In particular, once the selected commissioner role was achieved and the neighbouring municipalities and other potentially interested public actors had been excluded, the Municipality of Cabras seems to have had to incorporate the various demands of the private stakeholders in the implementation phase and redesign the interdependence map, at least in part, with a consequent further lowering in the degree of an explicit definition of the interdependence map within the policy network.

The evaluation of the degree of the second analytical dimension of the independent variable, namely, the (b) congruence of cognitive maps within the dominant coalition, can be considered *low* (Table 4.2). Our analysis shows the lack of distinction between technical and political decision-making, with an increase in bureaucratic burden and a scarce capacity for communication between the actors involved in the day-to-day management of the MPA. The central role in the construction and control of the cognitive maps in the coalition fell to the Ministry of the Environment, both in terms of establishing the area and in relation to all of the policy management phases where there was a lack of true subsidiarity in terms of the role of the Municipality of Cabras, apart from its formal role as selected commissioner. This top–down-type hierarchical structure generated a pronounced slowdown in the realisation of changes requested by the municipality, associations, interest groups and finally (although not finally in terms of importance) the population, during the policy reformulation and implementation phase and/or in terms of proposing and bringing about innovation.

To summarise what emerged in terms of the third analytical dimension of the independent variable (Table 4.2), the (c) structuring and impermeability of gatekeeping, the case study under analysis presents a *medium/low* degree. This assessment is based on the fact that while the control of resources was governed at the local level by public institutions like the municipality and region, in the reformulation and implementation phase, the same actors were not able to exercise their functions in a completely autonomous manner concerning the sectoral and micro-sectoral interests, within a system which tended to be permeable. In particular, the critique noted in this area relates to the typically distributive elements of the policy, which allowed actors external to the dominant coalition, such as pressure and interest groups, committees, representatives of unions and environmental groups to access the resources at stake easily.

All this considered, as shown in the table above (Table 4.2), the overall level of institutionalisation of the governance system in Sinis peninsula – Mal di Ventre island MPA can be assessed as *medium/low*.

Table 4.3 Overall evaluation of the level of quality of environmental protection policy in Sinis peninsula – Mal di Ventre island MPA

Analytical dimensions of the quality of environmental protection policy	Overall evaluation in Sinis peninsula – Mal di Ventre island MPA
(a) Degree of responsiveness	Medium/low
(b) Degree of cooperation	Low
(c) Degree of functional congruence	Medium/low
Level of quality of environmental protection policy	
Medium/low	

4.5.2 Evaluation of the Quality of Environmental Protection Policy

As far as concerns the first analytical dimension of the dependent variable —namely, the (a) degree of responsiveness—we can affirm, as shown in the table below (Table 4.3), that the Sinis peninsula – Mal di Ventre island MPA case attests to the presence of a *medium/low* degree of policy response capability, in terms of the demands of the general population. This is because of the inadequacy of the results achieved in terms of sustainable fishing objectives, improvement in the safeguarding and valorisation of the biological and geomorphological resources of the area and replenishment of fish stocks and the launching of the local tourist industry. The establishment of the MPA, in fact, did not lead, as anticipated by the legislator, to the valorisation of the ecological, archaeological and cultural heritage of the area, and there was little or no impact in terms of local infrastructural development.

Moving on to an evaluation of the second analytical dimension of the dependent variable, that is, the (b) degree of cooperation, it is possible to rate a *low* degree of cooperation between the actors in the policy network, above all in the policy implementation phase (Table 4.3). The Municipality of Cabras, as the selected commissioner, revealed itself incapable of controlling processes with its resources or managing/coordinating actors superordinate to itself in the hierarchy and functionally independent and which were fundamental for the exercise of responsibilities essential to the correct functioning of the area (e.g. the enforcement of current MPA regulations). There was also a tendency towards conflict between the actors involved in the public–private, private–private and public–public fields.

Finally, the degree of the third and last analytical dimension of the dependent variable in this typical case study—namely, the (c) functional congruence—can be rated as *medium/low* (Table 4.3). The empirical analysis shows that, on the one hand, the dominant coalition seems to have succeeded in partially achieving objectives in terms of actors indirectly involved in the establishment of the MPA (in particular tourists or citizens) by guaranteeing a medium/high degree of functional congruence in terms of interventions. On the other hand, in terms of actors employed in a profession directly linked to the exploitation of the marine environment and its resources (mainly fishers), the policy does not seem to have been able to modify their behaviour and achieved a low level of functional congruence. Rather, the imposition of regulations, associated with a lack of surveillance in their implementation, seems to have generated new conflicts and non-virtuous behaviours compared to the period before the introduction of the policy, within a repeated zero-sum game between the parties.

Based on all the above considerations, the overall level of quality of the environmental protection policy in the Sinis peninsula – Mal di Ventre island MPA can be assessed as *medium/low* (Table 4.3).

References

Abrami A (2000) Il regime giuridico delle aree protette. Giappichelli, Torino

Airoldi L, Beck MW (2007) Loss, status and trends for coastal marine habitats of Europe. Oceanography and marine biology. An Ann Rev 45:345–405

Baker S, Milton K, Yearley S (eds) (1994) Protecting the periphery: environmental policy in peripheral regions of the European Union. Routledge, London

Bodansky D (2016) The Paris climate change agreement: a new hope? Am J Int Law 110 (2):288–319

Boehmer-Christiansen S (1994a) Global climate protection policy: the limits of scientific advice: part 1. Glob Environ Chang 4(2):140–159

Boehmer-Christiansen S (1994b) Global climate protection policy: the limits of scientific advice: part 2. Glob Environ Chang 4(3):185–200

Bulsei GL (1990) Le politiche ambientali: intervento pubblico e regolazione sociale. Rosenberg & Sellier, Torino

Bulsei GL (2005) Ambiente e politiche pubbliche. Dai concetti ai percorsi di ricerca. Carocci, Roma

Calaresu M (2017) The top–down instruments for governing crime and disorder: what lessons can be drawn from the Italian experience (2007–2011). In: Peršak N (ed) Regulation and social control of incivilities. Routledge, New York, pp 35–54

Calaresu M, Selmini R (2017) Policing and urban control in Rome and Milan: a view from the southern edge of Europe. In: Devroe E, Edwards A, Ponsaers P (eds) Policing European metropolises. The politics of security in city–regions. Routledge, Abingdon, pp 182–200

Calaresu M, Tebaldi M (2015) Local security policies and the protection of territory: an analysis of the Italian experience (2007–2009). City Terr Archit 2(1):1–18

Cárcamo PF, Garay-Flühmann R, Gaymer CF (2014) Collaboration and knowledge networks in coastal resources management: how critical stakeholders interact for multiple-use marine protected area implementation. Ocean Coas Manag 91:5–16

Casola E, Lariccia M, Scardi M (2008) La pesca nelle aree marine protette italiane. Unimar, Roma

Casola E, Lariccia M, Scardi M (2014) Aree Marine Protette e pesca professionale. Unimar, Roma

Caveen AJ, Gray TS, Stead SM, Polunin NV (2013) MPA policy: what lies behind the science? Mar Policy 37:3–10

Caveen AJ, Polunin NV, Gray TS, Stead SM (2014) The controversy over marine protected areas: science meets policy. Springer, Heidelberg

Christie P, Lowry K, White AT, Oracion EG, Sievanen L, Pomeroy RS, Eisma RLV (2005) Key findings from a multidisciplinary examination of integrated coastal management process sustainability. Ocean Coast Manag 48(3–6):468–483

De Santo EM (2013) Missing marine protected area (MPA) targets: how the push for quantity over quality undermines sustainability and social justice. J Environ Manag 124:137–146

Dente B, Knoepfel P, Lewański R, Mannozzi S, Tozzi S (1984) Il controllo dell'inquinamento atmosferico in Italia: analisi di una politica regolativa. Officina, Roma

Dessler AE, Parson EA (2019) The science and politics of global climate change: a guide to the debate. Cambridge University Press, Cambridge

Edgar GJ, Stuart-Smith RD, Willis TJ, Kininmonth S, Baker SC (2014) Global conservation outcomes depend on marine protected areas with five key features. Nature 506:216–220

Fenberg PB, Caselle JE, Claudet J, Clemence M, Gaines SD (2012) The science of European marine re-serves: status, efficacy, and future needs. Mar Policy 36:1012–1021

Ferrajolo O (1999) Le aree specialmente protette del Mediterraneo. In: Marchisio S, Marchisio S, Della Fina V, Ferrajolo O, Salberini G, Tamburelli G (eds) Codice delle aree protette: trattati internazionali, norme comunitarie e legislazione italiana. Giuffrè, Milano

Garzia G (2011) Le aree marine protette: funzione amministrativa e nuovi strumenti per lo sviluppo sostenibile. Giuffrè, Milano

Guarnieri C (1989) Burocrazia e magistratura. In: Morlino L (ed) Scienza politica. Edizioni della Fondazione Agnelli, Torino, pp 223–240

Guidetti P, Milazzo M, Bussotti S, Molinari A, Murenu M (2008) Italian marine reserve effectiveness: does enforcement matter? Biol Conserv 141:699–709

Haas PM (1989) Do regimes matter? Epistemic communities and Mediterranean pollution control. Int Organ 43(3):377–403

Haas PM, Keohane RO, Levy MA, Gasser L (eds) (1993) Institutions for the earth: sources of effective international environmental protection. MIT Press, Cambridge

Jordan A (ed) (2005) Environmental policy in the EU: actors, institutions and processes. Routledge, London

Kannen A (2014) Challenges for marine spatial planning in the context of multiple sea uses, policy arenas and actors based on experiences from the German North Sea. Reg Environ Chang 14 (6):2139–2150

Katsanevakis S, Levin N, Coll M, Giakoumi S, Shkedi D (2015) Marine conservation challenges in an era of economic crisis and geopolitical instability: the case of the Mediterranean Sea. Mar Pollut 51:31–39

Lanzalaco L (1995) Istituzioni, organizzazioni, potere: introduzione all'analisi istituzionale della politica. La Nuova Italia Scientifica, Roma

Leenhardt P, Cazalet B, Salvat B, Claudet J, Feral F (2013) The rise of large-scale marine protected areas: conservation or geopolitics? Ocean Coast Manag 85:112–118

Leroy P, Crabb A (eds) (2012) The handbook of environmental policy evaluation. Earthscan, London

Lester JP (ed) (1995) Environmental politics and policy: theories and evidence. Duke University Press, Durham/London

Levin K, Cashore B, Bernstein S, Auld G (2012) Overcoming the tragedy of super wicked problems: constraining our future selves to ameliorate global climate change. Policy Sci 45 (2):123–152

Lewański RJ (1990) La politica ambientale. In: Dente B (ed) Le politiche pubbliche in Italia. Il Mulino, Bologna, pp 291–314

Lewański RJ (1997) Governare l'ambiente, attori e processi della politica ambientale. Il Mulino, Bologna

Lizzi R (1995) Politiche e processi regolativi. Una riflessione sugli studi comparati della regolazione ambientale. Rivista di Scienza dell'Amministrazione 2:73–130

Lowi TJ (1964) American business, public policy, case-studies, and political theory. World Polit 16(4):677–715

Lowi TJ (1972) Four system of policy, politics, and choice. Public Adm Rev 32(4):298–310

Manca Cossu M (1990) I pescatori di Cabras. Mezzi e tecniche della pesca e organizzazione della produzione. Editrice S'Alvure, Oristano

Marino D (ed) (2011) Le aree marine protette italiane. Stato, politiche, governance. Angeli, Milano

Marra S, Coppa S, Camedda A, Massaro G, de Lucia GA (2017) The exploitation of limpets in a Mediterranean marine protected area: assessing the effectiveness of protection in the intertidal zone. Mediterr Mar Sci 18(3):406–423

Massaro G (2010) Metodi di gestione delle aree marine protette: attività di ricerca per l'applicazione di tecniche ed indicatori di efficacia. Università Ca' Foscari di Venezia: Tesi di dottorato di ricerca in scienze ambientali

Owens S (2010) Learning across levels of governance: expert advice and the adoption of carbon dioxide emissions reduction targets in the UK. Glob Environ Chang 20(3):394–401

Potts T, Burdon D, Jackson E, Atkins J, Saunders J, Hastings E, Langmead O (2014) Do marine protected areas deliver flows of ecosystem services to support human welfare? Mar Policy 44:139–148

Regonini G (2001) Capire le politiche pubbliche. Bologna, Il Mulino

Stoutenborough JW (2020) Public opinion and the environment: an evolving literature. In: Konisky DM (ed) Handbook of US environmental policy. Edward Elgar Publishing, Cheltenham, pp 156–169

Tebaldi M, Calaresu M (2009) Valutare la democrazia: introduzione all'analisi della qualità democratica. Aracne, Roma

Urquhart J, Acott T, Zhao M (2013) Introduction: social and cultural impacts of marine fisheries. Mar Policy 37:1–2

Van Slobbe E, de Vriend HJ, Aarninkhof S, Lulofs K, de Vries M, Dircke P (2013) Building with nature: in search of resilient storm surge protection strategies. Nat Hazards 66(3):1461–1480

Weible CM, Sabatier PA (2005) Comparing policy networks: marine protected areas in California. Policy Stud J 33(2):181–201

Worm B, Barbier EB, Beaumont N, Duffy JE, Folke C (2006) Impacts of biodiversity loss on ocean eco-system services. Science 314:787–790

Yearley S, Baker S, Milton K (1994) Environmental policy and peripheral regions of the European Union: an introduction. Reg Fed Stud 4(1):1–21

Chapter 5
The Quality of Territorial Policies in Europe's Periphery: Lessons from Three Typical Case Studies in Sardinia

Abstract This chapter aims to generalise the empirical evidence of the research. Starting with a study of the policy areas of urban regeneration and environmental protection in the three typical case studies in Sardinia, we discuss the descriptive and prescriptive implications for the local development of other peripheral/insular European contexts. Regarding the descriptive implications, in the first part of the chapter, we highlight the empirical findings of the research based on the degree of institutionalisation of the governance system and the quality of territorial policy for each policy area. Regarding the prescriptive implications, in the second part of the chapter, we identify the best practices for territorial governance to improve the chances of local development in Europe's periphery.

Keywords Quality of territorial policies · Best practices · Europe's periphery

5.1 Empirical Findings and Theoretical Implications of the Research

Many scholars agree that peripheral areas are affected by territorial disadvantages, such as fragility, marginality and weak local development, compared with central areas (Danson and De Souza 2012; Komornichy and Czapiewski 2006; Labrianidis 2006). Such territorial disadvantages also apply to islands, as the peripheral areas that are the subject of the theoretical and empirical research conducted in this volume. Our research was, in fact, concerned with *if* and *how* the local development of islands, as peripheral areas, can be improved through territorial policies. In particular, we tried to understand (a) what political and institutional factors can be identified as specific variables which could explain the production of good quality territorial policies and (b) if it is possible to develop a theoretical pathway and a subsequent empirical test that specifies which features should be included in territorial policies in order to ensure that they are considered as good quality policies.

To conduct this study, we analysed the local context of the Italian island of Sardinia. More specifically, this research focused on the issues raised by scholars

© Springer Nature Switzerland AG 2021

M. Tebaldi, M. Calaresu, *The Quality of Territorial Policies in Europe's Periphery*, Urban and Landscape Perspectives 22,
https://doi.org/10.1007/978-3-030-62647-1_5

who suggest that public policies aimed at changing the weak growth conditions of peripheral areas—particularly in the southern Italian regions—did not always achieve their objective (Governa 2008; La Spina 2008). According to these researchers, such policies often failed, resulting in protests and complaints about the inefficient and discriminatory use of public resources by the political actors involved in territorial governance processes.

There is a growing consensus among scientific circles, public administrators, stakeholders and even citizens that there is a need to determine, based on solid scientific knowledge and empirical evidence, which factors are more effective than others for generating successful territorial policies. In particular, good quality policies should (a) meet the needs of the community, possibly improving existing social and economic conditions, (b) respond to the targeted social groups while protecting the functional consistency of the policy, and (c) promote cooperation between actors in the policy network, particularly during implementation of the policy, to enhance its effectiveness.

As we have already mentioned in the first chapter of the book (Chap. 1), we can state that the specific object of this research was the analysis of the political–institutional factors that we think have the greatest influence on the production of territorial policies and on the quality of such policies. Therefore, this study did not consider (except incidentally) long-term historical, political, economic and social factors that may influence the politico–institutional independent variables that were the focus of our analysis. Nor this study considered the concrete outcomes of the policies analysed in terms of local development.

The main assumption of our research was that the quality of a policy depends on the complex set of conditions that structure governance systems, including the presence of a certain degree of 'shadow of hierarchy' (Grote 2008; Héritier and Lehmkuhl 2008; Börzel and Risse 2010). This metaphorical term represents the set of institutional factors (rules, roles of authority, organisational and inter-organisational coordination tools) that can structure the governance system of territorial policies and make it more likely that good quality decisions will be made and implemented. Following a well-established approach in political and policy sciences, we used the term institutionalisation to refer to the process that leads to structural strengthening of the governance system of a policy. In our hypothesis, the main explanation for the quality of public policies (a *dependent* variable) from both descriptive and prescriptive perspectives is—ceteris paribus—the institutionalisation of a series of attributes of the government system or the governance system of the policy (an *independent* variable). A high level of institutionalisation tends to increase the quality of territorial policies. Vice versa, the quality of territorial policies tends to decrease with a low level of institutionalisation.

In order to empirically analyse the degree of institutionalisation of a governance system, we split the concept into three analytical dimensions: (1inst) *functional specialisation and interdependence maps among actors in the dominant coalition*, (2inst) *congruence of cognitive maps within the dominant coalition* and (3inst) *structuring and impermeability of gatekeeping*. The higher the degree of these

Table 5.1 Independent variable: institutionalisation of the governance system

Territorial policy	Case study	Analytical dimensions			Overall
		1inst	*2inst*	*3inst*	
Urban regeneration	*Sassari*	Medium	Medium	Medium/ low	Medium
Urban regeneration	*Alghero*	Medium/ high	Medium/ high	Medium/ low	Medium
Environmental protection	*Sinis*	Medium/ low	Low	Medium/ low	Medium/ low

Table 5.2 Dependent variable: quality of the territorial policies

Territorial policy	Case study	Analytical dimensions			Overall
		1qua	*2qua*	*3qua*	
Urban regeneration	*Sassari*	Low	Medium/ low	Medium/ low	Medium/ low
Urban regeneration	*Alghero*	Medium/ high	Medium/ high	Medium/ high	Medium/ high
Environmental protection	*Sinis*	Medium/ low	Low	Medium/ low	Medium/ low

three dimensions, the higher the level of institutionalisation of the governance system of the policy. Each dimension is assessed according to a five-level scale (i.e. low, medium/low, medium, medium/high, high).

We also split up the quality of public policy into three analytical dimensions: (1qua) *responsiveness*, (2qua) *cooperation* and (3qua) *functional congruence*. The higher the degree of these three dimensions, the higher the quality of a policy. Each dimension was assessed according to the same five-level scale. As a final step, we conducted an overall assessment of the two variables based on the average values of the three analytical dimensions for each concept on the five-level scale.

To understand whether our above-mentioned hypothesis could actually lead to a better understanding of the operational mechanisms of territorial policies and improve their performance, we tested it within three typical cases on the island of Sardinia from the early 1980s to the second half of the 2000s. Two case studies were territorial policies focusing on the urban regeneration processes carried out in the Old Towns of Sassari and Alghero between 1995 and 2006. The third case study was an environmental protection policy focusing mainly on the MPA of the Sinis peninsula – Mal di Ventre island, established within the Municipality of Cabras, from 1993 to 2004.

The research results are summarised in the following tables (Tables 5.1 and 5.2), which provide an overall assessment of the two variables, based on the average values of the three analytical dimensions, on the above-mentioned five-level scale.

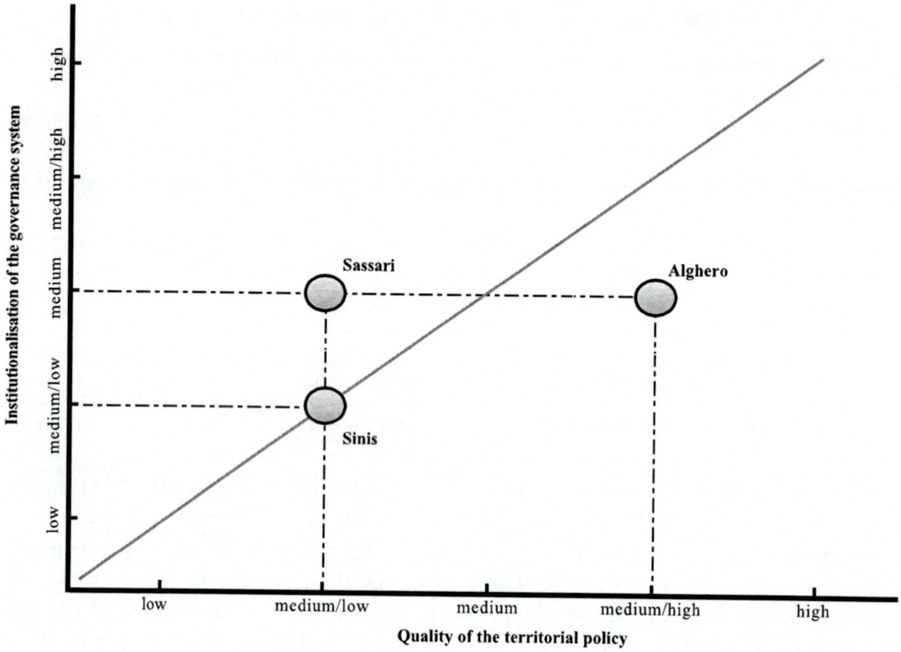

Fig. 5.1 Empirical results and the unified policy-oriented theoretical framework of the research

In the following figure (Fig. 5.1), we drew the intersections of the two variables on a Cartesian plane, with values for the independent variable on the y-axis and those for the dependent variable on the x-axis.

To empirically confirm the initial hypothesis, the overall values of the two variables should intersect along the straight line representing the correlation function. As the figure shows, only the case study relating to the Sinis peninsula – Mal di Ventre island environmental protection policy fully confirmed our hypothesis. The two urban regeneration policy cases showed a slight divergence between the intersection of the values assigned to the two variables and the straight line representing their theoretical position.

On the one hand, in the Sassari urban regeneration case, the overall degree of institutionalisation of the governance system corresponded to the overall medium–low quality of the policy. On the other hand, in the Alghero urban regeneration case, the overall degree of institutionalisation of the governance system indicated that the policy was of medium–high quality. Similar degrees of average institutionalisation of the governance system generated different results regarding the quality of the urban regeneration territorial policies; medium–low quality was reported for Sassari, while medium–high quality was reported for Alghero.

These results helped to focus on our theoretical framework. Overall, the hypothesis of a causal link between the independent variable (the degree of institutionalisation of the governance system) and the dependent variable (the quality of territorial policy) seemed to be confirmed, although further analysis is needed to fully explain the partial divergence between the theoretical framework and the results of our empirical analysis. This divergence showed that the correlation had very sound linearity for low values but was less linear when the variables reached medium values. In other words, empirical evidence suggested that a low or medium–low degree of the institutionalisation of the governance system tends to be associated with low or medium–low quality of the territorial policy. The accuracy of this correlation tends to decrease when the independent variable reaches a medium value.

Two factors can explain this result. The first factor relates to the methodology used to test the overall evaluation of both the dependent and independent variables in the three case studies. Our qualitative method estimated the mean values reported for the three dimensions of both variables, considering each dimension to be of equal weight in the analysis to the other two. However, the latter assumption may prove to be misleading and hide greater underlying complexity. If this is true, the three analytical dimensions of the independent and dependent variables should be considered to have different weights to estimate their overall value correctly. The second factor relates to the influence of intervening variables, which could slightly influence how the degree of institutionalisation of the governance system affects the quality of territorial policies.

With the term 'intervening variables', we refer to those factors that have limited influence but may change the cause–effect relationship between the main variables, such as economic factors (periods of local economic crisis, sudden changes in prices within local markets, etc.), technology-related factors (innovations that facilitate the implementation of a policy or suddenly make adopted solutions obsolete), social factors (the increase or decrease of migration flow at a local level) and weather- and climate-related factors (extreme and localised bad weather conditions).

Nevertheless, we can state that our study adds value to the current debate on the institutionalisation of the governance system and the quality of territorial policies. As far as the theoretical level of the debate is concerned, on the basis of our thick description of three selected case studies in an EU peripheral area (the island of Sardinia), it was possible to substantially confirm the hypothesis concerning the causal mechanisms operating between the two above-mentioned theoretical concepts. Our study confirms that hypothesis, offering empirical evidence and theoretical robustness to the thesis of the authors (Jessop 1997, 2000; Imrie and Raco 1999; Whitehead 2003) who have problematised 'the dualistic or binary representation of governance as a movement away from government (at a local, national and international level), [and have] consequently stressed the hybrid nature of governance (i.e. government + governance) and the persistence of hierarchical rule within governance relations' (Whitehead 2003, 6).

5.2 Limitations of the Work and Future Research Agenda

The empirical analysis of the three typical case studies revealed the theoretical potential of our approach and some limitations, which are essentially methodological in nature, and avenues for future research in this field of study. The limitations are the arbitrary assumptions we made: (a) that the analytical dimensions of both the independent and dependent variables were equivalent to each other and (b) that there were no intervening variables at the territorial level that may affect the cause–effect relationship of the two main variables.

The first limitation may be overcome by identifying a path for refinement of the method used in this research. This method should be able to take into account the specific weight of each analytical dimension for the assessment of the two variables. Thus, the overall calculation should no longer be derived from a simple average of the values of the analytical dimensions, but from a weighted average. To establish the reciprocal weights of the analytical dimensions and define the weighting criteria, future research may use several qualitative methods, such as the expert survey (Mair and Castles 1977), the Delphi method (Landeta 2006) or the elite interview (Richards 1996).

The second limitation may be overcome by testing the hypothesis using a different methodology from the one we adopted. The research design developed for this research is based on a typical case study strategy, which focuses on the selection of one or more units (i.e. cases) that represent the average relationship between an independent variable and a dependent variable within the range of available cases. This allowed us to test the hypotheses through an intensive analysis of the selected cases. Although exploring the relationship between the two main variables in depth may be considered an appropriate methodological strategy, it may overshadow, or even fail to detect, the effects of any intervening variables at the territorial level. To determine whether other relevant variables can affect the main hypothesis, future research may benefit from adopting the exemplary case study strategy (Seawright and Gerring 2008). Usually, exemplary case analysis is used to test other variables that may affect the relationship between an independent and dependent variable. Therefore, exemplary case analysis can provide clues about the presence of any missing variables (Lanzalaco and Prontera 2012, 57).

Moreover, when the qualitative method is used, the 'pitfalls' of drawing conclusions mostly based on elite interviews (Dexter 1970, 100) must be taken into account when assessing the level of institutionalisation. Hence, future research should extend the comparison to more case studies in the EU's periphery (e.g. EU islands that like Sardinia have more than 500,000 inhabitants, such as Sicily, Mallorca, Cyprus and Crete), the purpose being to subject the hypotheses of causal relationships to quantitative comparative analysis.

Based on these considerations and remaining aware of the methodological limitations of this work, it is worth noting that our empirical analysis confirmed, rather than disproved, the main initial hypothesis. Only one divergence was detected in the empirical results for the Sassari urban regeneration case study: a medium degree of institutionalisation of the governance system was associated with medium–low quality. However, the low value of the dependent variable had a minor (one-level) divergence from the value that would have fully confirmed our hypothesis. While this result seems to reverse the direction of the cause–effect relationship, it can be considered a minor divergence, and it should be revised according to the methodological adjustments described above and the weight of other potentially intervening variables. In the other two typical cases, the causal direction of the relationship between the variables was confirmed. In the urban regeneration case in Alghero, the degree of institutionalisation of the governance system (medium) had a more than a proportional capacity to affect the quality of territorial policy positively (medium–high). Two of the three case studies (Sinis peninsula – Mal di Ventre island and Alghero) confirmed the assumptions discussed in other comparative works, which used a similar theoretical framework to ours, based on the relationship between the institutionalisation of local governance arrangements and the performance of territorial policies.

Those works aim to contribute to the literature on territorial governance, starting from the hypothesis that governance systems can improve public policies (or, as we stated here, the quality of territorial policies), 'but only if policy makers have the option to hierarchically impose the policy, i.e. have the capacity to adopt and enforce it without the involvement and against the opposition of private actors' (Börzel et al. 2008, 2). This applies to any governance system, including those characterised by the non-hierarchical involvement of non-governmental actors. As Börzel and Risse argue (2010, 116), 'it is frequently overlooked in the literature that non-hierarchical modes of coordination, such as negotiation and competition systems, are often embedded in hierarchical structures'. The theoretical framework of the above-mentioned works is well described by Jessop, who argues that governments:

> [. . .] provide the ground rules for governance and the regulatory order in and through which governance partners can pursue their aims; ensure the compatibility or coherence of different governance mechanisms and regimes; act as the primary organizer of the dialogue among policy communities; deploy a relative monopoly of organizational intelligence and information with which to shape cognitive expectations; serve as a 'court of appeal' for disputes arising within and over governance; seek to re-balance power differentials by strengthening weaker forces or systems in the interests of system integration and/or social cohesion; try to modify the self-understanding of identities, strategic capacities, and interests of individual and collective actors in different strategic contexts and hence alter their implications for preferred strategies and tactics; and also assume political responsibility in the event of governance failure. This emerging role means that networking, negotiation, noise reduction, and negative as well as positive coordination occur 'in the shadow of hierarchy. (Jessop 2000, 25)

The same theoretical approach serves as the core of the research programme developed by Adrienne Héritier, which is described in detail by a special issue of the *Journal of Public Policy*, which she edited in 2008 with Dirk Lehmkuhl. The issue began by introducing the following research questions: 'is the shadow of hierarchy needed to guarantee the efficacy of policymaking and policy performance under sectoral governance? Which link between government and governance – along the specified dimensions – ensures a high policy performance?' (Héritier and Lehmkuhl 2008, 3).

During the first decade of the 2000s, the research agenda regarding the territorial governance of local development policies was enriched with a new, topical issue that had 'to do with statehood or, more precisely, with the limited statehood of local government and could be cast in terms of a "shadow of hierarchy"' (Grote 2008, 15). As we discussed above, in addition to having a strong tradition of research in political science and organisational theory, the concept of institutionalisation we used in our research allowed us to develop a policy-oriented theoretical framework for the study of governance systems. Furthermore, we developed a hypothesis regarding the cause–effect relationship in which the degree of institutionalisation of the governance system explains the quality of territorial policies. As already mentioned in this section, our initial hypothesis was empirically confirmed in at least two of the three Sardinian typical case studies. The results can be used to create advice for policy-makers aiming to develop territorial policies that have a good chance of successfully improving the development of peripheral areas.

5.3 Best Practices for Improving the Quality of Territorial Policies in Europe's Periphery

Empirical analysis showed a tendentially direct cause–effect relationship between the degree of the institutionalisation of the governance system and the quality of territorial policies. The three typical case studies also identified the most effective local governance arrangements that could be used by policy- and decision-makers to influence the performance of territorial policies.

The two cases of urban regeneration policies in the cities of Sassari and Alghero allowed us to evaluate some of the procedures for interaction between political and administrative institutions at different levels of government. It was particularly relevant to allow the policy-making process to proceed quickly throughout all stages of development and to achieve results of overall satisfactory quality. In both cases, our empirical findings revealed that the key resource around which actors in governance systems build opportunities to design and implement an urban regeneration process is the funding that the EU makes available to peripheral areas to support their development. The EU's funding opportunities sustain innovative local governance arrangements and trigger different configurations of the dominant coalition. In this perspective, we agree with Elisei (2011, 42), who wrote that 'structural funds are an opportunity to change old governance mechanisms and promote new and innovative forms of local governance'.

The institutionalisation of these innovative governance systems tends to have a significant effect on the policy-making process. Due to observance of the agreements between political and administrative institutions, the cohesion and robustness of the governance systems, sequential steps and different phases of the policy process are reviewed and modelled in order to rationalise and simplify public decision-making. With further details, the Sassari case study showed that since the formulation phase of the policy, the creation of a solidly linked administrative structure (a *steering committee*) between the Municipality of Sassari and the central government stimulated a positive impact on the actual development of urban regeneration process.

Evidence showed that the decision-making process speeds up when a 'policy entrepreneur' steers the process (in this case, the local government in charge of the city of Sassari between 1995 and 2000), which triggers a continuous functional relationship between the different governmental levels that temporarily bear the burden of decision-making. The Sassari case also demonstrated that the proactive role of the policy entrepreneur need not necessarily be implemented by the intermediate level of the territorial institution (in this case, the Region of Sardinia), in compliance with a hypothetical balancing based on strict inter-institutional hierarchies. In accordance with the principle of subsidiarity, which is well applied in this case, the process of building the dominant coalition—and, therefore, the core of the governance system—can be successfully prompted by a lower territorial level (i.e. the Municipality of Sassari), according to an effective bottom–up approach.

The Sassari case study clearly showed that the process of institutionalisation of the governance system is not necessarily intended as a result of a formal organisational activity resulting from the simultaneous action and participation of all actors within the policy network. In fact, the process of institutionalisation can proceed gradually, starting from the basis of bottom–up relationships between political and administrative structures with low or very low institutional, functional and territorial contiguity. This first nucleus of relationships tends to act as a 'great attractor' for the other institutional components of the policy network. The Sassari urban regeneration policy case study showed that the creation of a steering committee dramatically simplified and sped up the decision-making process. The policy-making cycle started directly from the decision-making phase, compressing the two first stages—construction of the political agenda and identification of policy alternatives—into a single fast-track phase and identifying the best practices for decision-makers.

As far as the Alghero case study is concerned, the empirical data suggested that the creation of implementation structures and local governance arrangements facilitate inter-institutional coordination at the territorial level during the decision-making process. As described in the case study analysis, every stage of the decision-making process was characterised by the use of collaborative methods and consultation as well as a precise definition of the competences of each actor involved in the policy network. The use of that method for the resolution of arguments also led to an absence of conflicts among the actors involved. Although the actors pursued divergent interests, the dominant coalition of the policy network successfully reached an agreement in almost all circumstances, bringing different interests together. More specifically, the Alghero case study demonstrated how the governance system managed complex urban regeneration policy process during the implementation phase.

The procedures developed by the dominant coalition of the governance system managed to establish a stable dialogue through communication channels—which were not always formalised, nevertheless were always institutionalised through daily practice—between the city government, its bureaucratic apparatuses and the technical and enforcement institutions controlled by the central government (superintendencies). In this regard, issues concerning the authorisation and execution of the actions taken by each political and administrative actor were overcome through a series of unwritten rules based on the principles of consultation and sharing among different subjects.

Regarding environmental protection policy in the typical case of the Sinis peninsula – Mal di Ventre island MPA, the empirical evidence showed that in regulatory contexts characterised by innovation, interdependence and complexity (Lowi 1964, 1972), the institutional actor responsible for proposing the policy (Lewański 1997), namely, the institutional policy entrepreneur, is particularly relevant, especially during the policy design phase. The policy entrepreneur is important for structuring a strong policy network and controlling and/or monitoring all stages of the policy process. This is particularly true if the political actor responsible for proposing the policy seeks to achieve good quality territorial policies, which can improve pre-existing social conditions, affect the expected social target groups and avoid causing additional problems for other social groups and/or the environment.

In the Sinis peninsula – Mal di Ventre island MPA case study, the institutional policy entrepreneur (the Ministry of the Environment) formally identified the Municipality of Cabras as the only selected commissioner of the MPA but failed to substantially guarantee to the municipality that the formal principle of subsidiarity would be applied. In fact, the Municipality of Cabras had not appropriate bureaucratic–administrative tools, knowledge, technical and scientific skills, and human resources to be used during the policy implementation phase without always depending on the central government. The interviews indicated that the Ministry of Environment was not able to effectively play a *super partes* role in the intermediation of interests or to foster cooperation among the political and administrative actors at different governmental levels (e.g. the municipalities and the region of Sardinia) within a policy network that was intentionally built to be state-centric (Tebaldi and Calaresu 2009).

The empirical analysis, in this case suggested two things. First, in such regulatory contexts, if the institutional policy entrepreneur chooses to build a top–down hierarchical network, the distribution of roles and functional responsibilities among the main actors of the dominant coalition should be defined as clearly as possible at the very beginning of the policy cycle. Second, it is important for the state-centric policy network to separate technical decisions (which should be made by peripheral governmental levels without a continuous need for consultation and approval by the central level) from political decisions, as this is an effective governance arrangement that can have a positive impact on the degree of institutionalisation of the governance system. Consequently, if policy-makers wish to favour territorial policies of good quality as a way to optimise and simplify public decisions, the political–administrative actors at various territorial levels of the dominant coalition who are empowered by the central government to implement territorial policies (in this case, the Municipality of Cabras as the selected commissioner of the MPA) should not encounter difficulties (or only few difficulties) in circumscribing the nature and limits of their competences within the policy network.

In line with what we highlighted above for the Sassari and Alghero case studies, the empirical evidence from the Sinis peninsula – Mal di Ventre island MPA case demonstrated that the main reason why the Municipality of Cabras wanted to play a leadership role in the dominant coalition was the chance to obtain major economic resources allocated by the Ministry of the Environment through EU funding rather than a specific interest in adopting environmental protection policy within its own territory. However, unlike the Sassari and Alghero case studies, the potential availability of new financial and economic resources resulting from the establishment of the marine area did not prompt particularly innovative local governance arrangements in Sinis peninsula – Mal di Ventre island MPA. Instead, the low degree of gatekeeping of the dominant coalition generated fragmented, disorganised programmes and actions.

Moreover, the decision-making process and its implementation were affected by specific constellations of groups seeking an appropriation of disaggregated resources. A possible explanation for this is the desire of the Municipality of Cabras to exclusively acquire the potentially available resources for the creation and management of the MPA (rather than achieve shared allocation with several territorial entities or handle funding management through an intermediate body, such as the province or region). In addition, as the only selected commissioner for the MPA, the municipality may have wanted to obtain a full and unchallenged ability to make decisions about how those resources should be used once acquired.

Another possible explanation for the lack of new differentiated structures and procedures in the case of Sinis peninsula – Mal di Ventre island MPA (which is not mutually exclusive from the previous one) could be that at the time of policy adoption, local political and administrative elites were concerned that the Ministry of Environment could have opted for a sociocentric policy network and thus empowered or entrusted environmentalist associations with the management of the MPA. This had happened in other marine protected areas of Italy. Consequently, the MPA could have burdened by the maximum level of environmental, landscape and naturalistic constraints envisaged by national law, affecting the tourism and fishing sectors and 'overprotecting' the environment. To prevent this worst-case scenario, the city mayor and city council of Cabras chose to handle the MPA on their own, pushing out local lobbyists and active interest groups (e.g. boaters, fishers, tourists and citizens) from the dominant coalition since the very first phases of the policy cycle. This also happened to other local political–administrative institutions, such as the neighbouring municipalities and the law enforcement agencies, responsible for ensuring compliance with the rules for the protected area, including the port authority.

We, therefore, highlight the importance for institutional actors at peripheral levels of government—for achieving good quality policy in a state-centric policy network—to sustain, from the earliest stages of the policy cycle, innovative local governance arrangements (e.g. a formal institutional coordination group and/or round tables, concertation tables and steering committees) that are able to include different political–administrative and non-institutional actors in institutionalised communication. It is also important, given the complexity and variety of interests involved in the decision-making process, the lack of creation of such structures and the mechanisms in the Sinis peninsula – Mal di Ventre island MPA case did not facilitate inter-institutional coordination during the decision-making process, preventing policy-makers from considering both environmental and social/economic needs.

As described in the case study analysis, every stage of the decision-making process was characterised by continuous requests for revising the parameters and rules for the management of the MPA just 2 years after its formal approval. Additionally, there was a tendency for a high level of conflict between involved actors who pursued (potentially) divergent interests in the policy network on the public–private axis (e.g. the city mayor and fishers' trade unions), private–private axis (e.g. environmental associations and tourists) and public–public axis (e.g. the port authority and other public–administrative bodies involved in the management of the MPA). As the Sinis peninsula – Mal di Ventre island MPA case study showed, due to the lack of use of collaborative methods, consultations and precise definitions of the competences of each actor involved in the policy network, the dominant coalition did not succeed in finding a common goal and/or resolution that brought together the different interests of people working in economic sectors related to the use of the marine environment and its resources, such as fishers and tourists. As illustrated by the Sinis peninsula – Mal di Ventre island MPA typical case study, especially during the implementation phase, the low degree of institutionalisation of the governance system led to exogenous pressures by small groups of stakeholders who declared that their economic and social conditions and the status of their environment were even worse than before the policy was adopted.

Our examination of the above-mentioned case also demonstrated that with a medium–low degree of institutionalisation of the governance system, top–down imposition of a regulatory setting for management of the MPA led to a poor understanding of the objectives of the policy and a lack of internalisation, especially by non-institutional stakeholders. In particular, the lack of internalisation in the Sinis peninsula – Mal di Ventre island MPA was associated with a failure or at least with discontinuous and biased surveillance by the law enforcement agencies responsible for ensuring compliance with the rules for the MPA. Thus, conflicts and non-virtuous behaviours were generated among the actors outside the dominant coalition, and there was a need for constant renegotiation of the terms of agreements reached by the involved actors (e.g. the relationship between fishers' trade unions and national and local law enforcement agencies).

As a final lesson from this typical case study, it seems crucial for policy-makers to pursue a high level of institutionalisation of the governance system, as this would guarantee an increase in the quality of policy, particularly regarding its ability to respond to communities' needs and improve pre-existing social conditions, thus affecting the social target groups in peripheral areas.

References

Börzel TA, Risse T (2010) Governance without a state: can it work? Regul Gov 4(2):113–134

Börzel TA, Buzogany A, Guttenbrunner S (2008) New modes of governance in accession countries. The role of private actors. Paper presented at the NEWGOV Cross-Cluster Workshop "Civil society, new modes of governance and enlargement". Free University Berlin, https://ceses.cuni.cz/CESES-141-version1-6_2__Borzel_Buzogany_2008_CEE_Private_Actors_environment.pdf

Danson M, De Souza P (eds) (2012) Regional development in northern Europe: peripherality, marginality and border issues. Routledge, London

Dexter L (1970) Elite and specialised interviewing. Northwestern University Press, Evanston

Elisei P (2011) EU tools and urban regeneration. In: Mironowicz I, Ryser J (eds) Urban change. The prospect of transformation. Un-Habitat and Wroclaw University of Technology, Wroclaw, pp 40–43

Governa F (2008) Local development in fragile areas. Critical remarks on the initiatives underway in the mountains of Lombardy. J Alp Res 96(3):1–15

Grote JR (2008) Local governance and organized civil society: concepts and applications. CINEFOGO Conference on New Forms of Local Governance and Civil Society, Trento, https://ceses.cuni.cz/CESES-136-version1-5D_LGA_Grote_2008.pdf

Héritier A, Lehmkuhl D (2008) Introduction. The shadow of hierarchy and new modes of governance. J Publ Policy 28(1):1–17

Imrie R, Raco M (1999) How new is the new local governance? Lessons from the United Kingdom. Trans Inst Br Geogr 24(1):45–63

Jessop B (1997) Capitalism and its future: remarks on regulation, government and governance. Rev Int Polit Econ 4(3):561–581

Jessop B (2000) The dynamics of partnership and governance failure. In: Stoker G (ed) The new politics of local governance in Britain. Macmillan, Basingstoke, pp 11–32

Komornichy T, Czapiewski K (eds) (2006) Europa XXI. Regional periphery in central and eastern Europe. Stanislaw Leszczycki Institute of Geography and Spatial Organization, Warszawa

La Spina A (2008) Ambitions and failures of some local development policies for the Mezzogiorno. Territorial agreements and integrated territorial programs. Quad Sociol 48:21–38

Labrianidis L (2006) Fostering entrepreneurship as a means to overcome barriers to development of rural peripheral areas in Europe. Eur Plan Stud 14(1):3–8

Landeta J (2006) Current validity of the Delphi method in social sciences. Technol Forecast Soc Chang 73(5):467–482

Lanzalaco L, Prontera A (2012) Politiche pubbliche comparate. Il Mulino, Bologna

Lewański R (1997) Governare l'ambiente, attori e processi della politica ambientale. Il Mulino, Bologna

Lowi TJ (1964) American business, public policy, case-studies, and political theory. World Polit 16 (4):677–715

Lowi TJ (1972) Four systems of policy, politics, and choice. Public Adm Rev 32(4):298–310

Mair P, Castles F (1977) Reflections: revisiting expert judgements. Eur J Polit Res 31:150–157

Richards D (1996) Elite interviewing: approaches and pitfalls. Politics 16(3):199–204

Seawright J, Gerring J (2008) Case selection techniques in case study research: a menu of qualitative and quantitative options. Polit Res Q 61(2):294–308

Tebaldi M, Calaresu M (2009) Valutare la democrazia: introduzione all'analisi della qualità democratica. Aracne, Roma

Whitehead M (2003) 'In the shadow of hierarchy': meta-governance, policy reform and urban regeneration in the West Midlands. Area 35(1):6–14

Appendix

Timeframe of the research for the typical case studies of Sassari and Alghero:
From 1995 to 2006
Interviewing timeframe:
From June to September 2015
Duration of individual interviews:
1.5 hours

Timeframe of the research for the typical case study of Sinis peninsula – Mal di Ventre island MPA:
From 1993 to 2004
Interviewing timeframe:
From February to March 2013
Duration of individual interviews:
1.5 hours

Total interviews for the three typical case studies:
51

Data availability:
All interviews are available, upon request, in digital and printed versions in Italian language.

List of interviewees for the three typical case studies:

Group A: *Mayors* (Sindaci) *of the Municipalities of Alghero, Cabras and Sassari*

- Carlo Sechi (Mayor of the Municipality of Alghero, from 1994 to 1998)
- Tonino Baldino (Mayor of the Municipality of Alghero, from 1998 to 2002)

© Springer Nature Switzerland AG 2021
M. Tebaldi, M. Calaresu, *The Quality of Territorial Policies in Europe's Periphery*, Urban and Landscape Perspectives 22,
https://doi.org/10.1007/978-3-030-62647-1

- Gian Marco Tedde (Mayor of the Municipality of Alghero, from 2002 to 2007)
- Pasquale Castangia (Mayor of the Municipality of Cabras, from 1994 to 1998)
- Efisio Trincas (Mayor of the Municipality of Cabras, from 1998 to 2008)
- Anna Sanna (Mayor of the Municipality of Sassari, from 1995 to 2000)
- Gian Vittorio Campus (Mayor of the Municipality of Sassari, from 2000 to 2005)
- Gianfranco Ganau (Mayor of the Municipality of Sassari, from 2005 to 2010)

Group B: *City Councillors* (Assessori delle giunte comunali) *of the Municipalities of Alghero, Cabras and Sassari*

- Fulvio Dettori (City Councillor for Public Works of the Municipality of Alghero, from 1994 to 1998)
- Mario Conoci (City Councillor for Public Works of the Municipality of Alghero, from 1998 to 2002)
- Piero Canu (City Councillor for City Planning of the Municipality of Alghero, from 1998 to 2002)
- Giancarlo Piras (City Councillor for Public Works of the Municipality of Alghero, from 2002 to 2007)
- Alberto Deiala (City Councillor for the Environment of the Municipality of Cabras, from 1986 to 1990)
- Chiara Maria Sanna (City Councillor for the Environment of the Municipality of Cabras, from 1994 to 1998)
- Alessandro Murana (City Councillor for the Environment of the Municipality of Cabras, from 1998 to 2008)
- Luigi Lotto (City Councillor for Public Works of the Municipality of Sassari, from 1995 and 2000 and from 2005 to 2009)
- Ottaviano Canalis (City Councillor for Public Works of the Municipality of Sassari, from 2000 to 2005)

Group C: *Members of the Local Councils* (Consiglieri comunali) *of the Municipalities of Alghero and Sassari*

- Angelo Caria (Local Councillor of the Municipality of Alghero, from 1994 to 1998)
- Elias Vacca (Local Councillor of the Municipality of Alghero, from 1998 to 2002)
- Giuseppe Tilloca (Local Councillor of the Municipality of Alghero, from 2002 to 2007)
- Nanni Moro (Local Councillor of the Municipality of Sassari, from 1995 to 2000)

- Luciano Chessa (Local Councillor of the Municipality of Sassari, from 2000 to 2005)
- Antonello Desole (Local Councillor of the Municipality of Sassari, from 2005 to 2010)

Group D: *Executive Officers* (Dirigenti) *and Officers* (Funzionari) *of the Municipalities and Soprintendenzas of Alghero and Sassari and Sinis peninsula – Mal di Ventre Island MPA*

- Gian Marco Saba (Executive Officer of Public Works of the Municipality of Alghero, from 1995 to 2006)
- Ivan Meloni (Official of Public Works of the Municipality of Alghero, from 1998 to 2006)
- Giommaria Angioj (Official of Public Works of the Municipality of Alghero, from 1994 to 2006)
- Ingrid Crabuzza (Official of Public Works of the Municipality of Alghero, from 1994 to 2006)
- Salvatore Chelo (Official of Public Works of the Municipality of Alghero, from 2000 to 2006)
- Salvatore Masia (Executive Officer of Planning and Control of the Municipality of Alghero, from 2000 to 2006)
- Bruno Paliaga (Director of the Sinis Peninsula – Mal di Ventre Island MPA, from 2001 to 2009)
- Giancarlo Budroni (Executive Officer of Public Works of the Municipality of Sassari, from 1995 to 2003)
- Salvatore Mongili (Official of Public Works of the Municipality of Sassari, from 1996 to 2006)
- Teodolinda Pilia (Official of Public Works of the Municipality of Sassari, from 1996 to 2006)
- Ida Rinaldi (Official of Public Works of the Municipality of Sassari, from 1997 to 2006)
- Claudio Castagna (Executive Officer of Public Works of the Municipality of Sassari, from 2003 to 2006)
- Ivan Meloni (Official of Public Works of the Municipality of Sassari, from 1998 to 2006)
- Daniela Scudino (Official of the Soprintendenza of Monumental Heritage, from 1995 to 2006)
- Daniela Rovina (Official of the Soprintendenza of Archaeological Heritage, from 1995 to 2006)
- Gianluca Zini (Official of the Soprintendenza of Monumental Heritage, from 1995 to 2006)

Group E: *Representatives of Stakeholders, Trade Unions and Environmental Associations (in Alphabetical Order)*

- Massimo Cadeddu (Confcommercio Nord Sardegna, from 1995 to 2006)
- Dario Cossu (WWF Seu, from 1987 to 1998)
- Antonio Cugia (Italia Nostra Sassari, from 1995 to 2006)
- Giovanni De Falco (International Marine Centre, from 1994 to 2006)
- Antonio Luigi Medardi (Federcoopesca Sardegna, since 1994)
- Walter Piras (WWF Oristano, from 1993 to 1997)
- Maurizio Porcu (WWF Oristano, from 2001 to 2003)
- Roberto Salmon (Italia Nostra Alghero, from 1995 to 2006)
- Gian Luigi Tolu (Confcommercio Sassari, from 1995 to 2006)
- Gianni Usai (Legacoop Oristano, from 1998 to 2003)
- Sebastiano Venneri (Legambiente, since 1991)
- Ivo Zoccheddu (Associazione Diportisti Nautici Cabras, from 1997 to 2002)

The interviews are based on the following semi-structured outlines (A–B) with an analytical grid as a benchmark:

(A) ***Questions to the Interviewees for the Typical Case Studies of Sassari and Alghero***

1. The 1990s in Italy were characterized by the inclusion of Old Towns' urban regeneration in the political agendas.

   1.1. Considering the political agenda, how do you think the problem was addressed, both centrally and peripherally?
   1.2. How do you think the issue should have been addressed?
   1.3. By using which instruments?

2. Urban regeneration policies during the 1994–1999 and 2000–2006 planning periods were an opportunity for development, and they could potentially bring about changes in the surrounding urban environment.

   2.1 In the case of Sassari/Alghero, do you think such policies achieved the expected results?
   2.2 How did territorial institutions relate to this opportunity?
   2.3 Could the administrations adequately support and implement political decisions?

3. Taking into account the years between 1995 and 2006, several municipalities of different political affiliation took turns in ruling the Municipality of Sassari/Alghero.

   3.1 Was there a common thread in the political intentions of the alternating ruling administrations over time?
   3.2 What were the reasons for any variations in the execution of public works?

3.3 What did and what should have the different ruling political majorities done to have a conversation with citizens and stakeholders?
3.4 Did the opinion of the stakeholders influence the choices made by the Municipality of Sassari/Alghero?

4. Urban regeneration policies aim at recovering the 'built' environments by restoring the existing building heritage, supporting historical–cultural and environmental resources available on the territory and improving the quality of life among the resident population.

4.1 Which instruments were used to secure the success of such policies?
4.2 How were those instruments adjusted over time?
4.3 What did the Municipality of Sassari/Alghero do, and what should it have done to achieve an effective process of regeneration within the Old Town?

5. In general, urban regeneration policies generate high costs that can hardly be fully sustained at the local level. For this reason, the EU has made a significant contribution to these policies through Structural Funds.

5.1 Considering the functioning of Structural Funds, were territorial institutions autonomous from decision-making and executive points of view?
5.2 If so, to what extent?
5.3 Considering such policies, could territorial institutions take action using their own resources?
5.4 If so, to what extent?

(B) ***Questions to the Interviewees for the Typical Case Study of Sinis Peninsula – Mal di Ventre Island***

1. In 1986, the Italian Ministry of the Environment, which had been recently replaced the Ministry of Merchant Navy, undertook the investigations provided by Law 979/1982 dedicated to the MPAs, also including the MPA of the Sinis peninsula – Mal di Ventre island.

1.1. Why did the Municipality of Cabras decide to undertake the establishment of the MPA?
1.2. At that time, could any other institutional body or association request the establishment and the management of the MPA to the Italian Ministry?
1.3. What role did the Ministry of the Environment play in the establishment of the MPA?

2. According to Italian rules and regulations, the MPA's environmental protection policy aims at preserving the characteristics of the area to be safeguarded, protecting and promoting its biological resources, carrying out scientific studies and research programmes and promoting a social and economic development that is compatible with the natural and landscape relevance of the area.

2.1 Which of the above were the most important outcomes for the management of the MPA?
2.2 Which objectives were and should have been taken into account mostly?

2.3 Which instruments were used to secure the achievement of such objectives?
2.4 How were those instruments adjusted over time?
2.5 How much did the opinion of stakeholders influence the drafting, discussion and implementation of the policy?
2.6 What role did experts and researchers play at the various stages of the policy cycle?

3. In conjunction with the establishment of the MPA, the Ministry of the Environment and the EU allocated considerable financial resources, also through Structural Funds. If used effectively, such resources were intended to boost employment and structural investment within the territory.

3.1 What role did financial resources play in setting up the MPA?
3.2 What needs were met using financial resources, and what needs should they have been used for?
3.3 Considering the functioning of Structural Funds, were the territorial institutions autonomous from decision-making and executive points of view?
3.4 If so, to what extent?
3.5 Considering such policies, could territorial institutions take action using their own resources?
3.6 If so, to what extent?

4. The decision to set up an MPA involved several categories of stakeholders: on the one hand, those who were directly interested in the implementation of the MPA rules, as they were employed in a profession linked to the exploitation of the marine environment and its resources, and on the other hand, those indirectly interested in its implementation, being potential users of the area.

4.1 What was done, and what should have been done to actively engage the people employed in sea-related activities and ensure a balance with the objectives of the MPA?
4.2 Did the opinion of the aforementioned directly or indirectly interested stakeholders influence any choice related to the MPA?
4.3 Did the establishment of the MPA influence the activities of the indirectly involved stakeholders?

5. Since its establishment, the MPA has undergone three changes by the national government through the issue of ministerial decrees, also following a request by the Municipality of Cabras.

5.1 What were the reasons for the Municipality of Cabras to request such changes?
5.2 Why did the following MPA zoning/boundary delimitation lead to a progressive decrease of the full protection Zone A?
5.3 Taking into account the MPA policy, was there a common thread in the political intentions of the alternating ruling administrations over time?
5.4 What are the reasons for the non-application or partial application of MPA regulations?